WITHDRAWN FROM LIBRARY

MONTGOMERY COLLEGE LIBRARY
ROCKVILLE CAMPUS

STUDIES IN REAL AND COMPLEX ANALYSIS

Studies in Mathematics

The Mathematical Association of America

Volume 1: STUDIES IN MODERN ANALYSIS
edited by R. C. Buck

Volume 2: STUDIES IN MODERN ALGEBRA
edited by A. A. Albert

Volume 3: STUDIES IN REAL AND COMPLEX ANALYSIS
edited by I. I. Hirschman, Jr.

Volume 4: STUDIES IN GLOBAL GEOMETRY AND ANALYSIS
edited by S. S. Chern

Volume 5: STUDIES IN MODERN TOPOLOGY
edited by P. J. Hilton

Volume 6: STUDIES IN NUMBER THEORY
edited by W. J. LeVeque

H. J. Bremermann
University of California, Berkeley

Lawrence M. Graves
University of Chicago and Illinois Institute of Technology

Einar Hille
Yale University

I. I. Hirschman, Jr.
Washington University

D. V. Widder
Harvard University

H. H. Schaefer
University of Tubingen

Guido Weiss
Washington University

Harold Widom
Cornell University

Studies in Mathematics

Volume 3

STUDIES IN REAL AND COMPLEX ANALYSIS

I. I. Hirschman, Jr., editor
Washington University

Published and Distributed by
The Mathematical Association of America

© *1965 by*
The Mathematical Association of America (Incorporated)
Library of Congress Catalog Card Number 65-22403

Complete Set ISBN 0-88385-100-8
Vol. 3 ISBN 0-88385-103-2

Printed in the United States of America

Current printing (last digit):

11 10 9 8 7 6 5 4 3

CONTENTS

INTRODUCTION

I. I. Hirschman, Jr.

The eight articles in the present volume do not all presuppose the same mathematical background; they are directed generally to readers at the advanced undergraduate and first-year graduate level.

The initial article by H. J. Bremermann is a description of part of the modern theory of several complex variables which is centered about the successful efforts of mathematicians to understand fully the remarkable continuation properties possessed by analytic functions of several complex variables. Other topics central in this theory, such as the Cousin problems, analytic sets, etc., are discussed, although more briefly.

Graves' paper deals with a less extensive area, that of nonlinear functions from one Banach space to another, and in particular with the implicit function theorem. The material considered is treated in detail. Since this subject is beginning to make its way into advanced calculus texts, it is particularly fortunate to have this exposition. It is to be noted that Graves' paper has some elements in common with "Preliminaries to Functional Analysis" by Casper Goffman in Volume 1 of this series and that the two papers can be profitably read together.

Hille's paper on semi-groups gives a brief description of this vast area of analysis. The reader is introduced to such central, structural features of semigroup theory as the resolvant and the infinitesimal generator, and is also afforded a hint of the applications of this theory to stochastic processes and partial differential equations. Hille's article also makes contact with that of Goffman referred to above.

The article written by Hirschman and Widder is devoted to a relatively specific problem—the genesis of the real inversion formulas of the Laplace and Stieltjes transforms. These formulas and the corresponding representation theory are now seen, after some decades, to be a partly autonomous chapter within the very extensive area of totally positive matrices, variation diminishing transformations, and their extensions and generalizations.

Schaefer's paper is entirely different in spirit than the others in this volume in that it treats a classical subject, the Lebesgue-Stieltjes integral, in detail. Schaefer's approach is that of Daniell and F. Riesz; that is, the Lebesgue integral is constructed by an extension process from the Riemann integral, the theory of measure appearing only as a byproduct and at the end. Because it is both brief and rather complete, Schaefer's paper affords a unique opportunity to sample the elegance of this less familiar method. Moreover, this paper can serve as a convenient source for many of the measure theoretic results required in the other papers of these volumes.

Weiss' paper is simultaneously a detailed exposition of certain basic parts of harmonic analyses and an introduction to and description of selected advanced topics. The principal emphasis is on harmonic analysis in its classical form and here the exposition introduces the reader to the concept of "weak type" and to the Marcinkiewicz interpolation theorem, ideas which have played an important role in harmonic analysis in the last decade. The article concludes with a brief discussion of abstract harmonic analysis on locally compact Abelian groups.

Widom's paper is addressed to a rather specific problem, the inversion of semi-infinite Toeplitz operators. It can be profitably read in conjunction with Lorch's "The Spectral Theorem" in Volume 1 of this series. It is particularly interesting to see how, confronted with a concrete problem in spectral theory, the author draws on other phases of mathematics, in this case on the theory of Fourier series and analytic functions, in order to obtain a solution.

The articles of this volume treat only a small sample from the many topics of current interest in analysis, but it is believed that they are an interesting selection and it is hoped that the present volume will be a worthy successor to the elegant "Studies in Modern Analysis," which is Volume 1 in this series.

SEVERAL COMPLEX VARIABLES

H. J. Bremermann

The theory had its beginning shortly before the turn of the century. At first concepts and methods of the theory of one complex variable were generalized. Very soon, however, problems were encountered that were well understood in the case of one variable, but defied solution for two and more variables. Also, F. Hartogs [26], [27] (between 1906 and 1910) discovered profound results about analytic continuation and "natural boundaries" that are false for one variable. It became clear that the theory of several complex variables is not a mere generalization from one to n, but a distinct theory of its own.

After Hartogs, progress was slow for about twenty years. Then H. Behnke, H. Cartan, and P. Thullen developed the theory of domains and envelopes of holomorphy and S. Bergman began to investigate the kernel function and invariant metric (called after him).

In 1934 Behnke and Thullen summarized the knowledge up to that time in their book, *Theorie der Funktionen mehrerer komplexer Veränderlichen* [2] (still of interest).

Some of the outstanding problems mentioned in Behnke-Thullen have since been solved: (1) the analogue of "Runge's theorem," (2) construction of a meromorphic function to locally given poles (the so-called "additive Cousin problem") and construction of a holomorphic function to locally given zeros (multiplicative Cousin problem), and (3) the local characterization of the domains of holomorphy. The solutions of these problems are mostly due to K. Oka [32]–[40].

In recent years investigations have proceeded to complex manifolds and lately to "complex spaces," which are the n-dimensional analogues of Riemann surfaces. The language of "sheaves" has been developed and found to be an appropriate and powerful tool for studying functions and sets of functions on manifolds and complex spaces. Also, connections with Banach algebras have developed, and recently several complex variables have become important in theoretical physics (quantum field theory) [16].

Recently several books and notes on several complex variables have become available. B. A. Fuks [21] has appeared in a new and completely revised edition (translated into English) and a second volume has been added [22]. Topics that are of importance for quantum field theory have been treated by Vladimirov [50]. Excellent lecture notes have been compiled by L. Bers [7]. L. Hörmander [28] solves problems (1), (2), and (3). Of an earlier date are: Bochner-Martin [3] and Cartan seminaire 1951–1952 [19].

In what follows I will try to give a glimpse of the theory by emphasizing the problems mentioned above, around which much of the research has grown.

It is impossible in this limited space to deal with "sheaves," "complex manifolds," and "complex spaces." We can only touch on these subjects, giving references to the original literature. We also had to leave out Bergman's theory. An excellent introduction to this theory can be found in Bergman [5, chap. 11], and a more detailed representation in Bergman [6].

1. THE SPACE OF n-TUPLES OF COMPLEX NUMBERS C^n

From the familiar complex numbers we may form n-tuples. The collection of all n-tuples $z = (z_1, \cdots, z_n)$ of complex numbers $z_1, \cdots, z_n$ is denoted by C^n. We make it a linear vector space by introducing addition

$$z^{(1)} + z^{(2)} = (z_1^{(1)}, \cdots, z_n^{(1)}) + (z_1^{(2)}, \cdots, z_n^{(2)})$$
$$= (z_1^{(1)} + z_1^{(2)}, \cdots, z_n^{(1)} + z_n^{(2)});$$

and multiplication with a complex scalar λ

$$\lambda z = \lambda(z_1, \cdots, z_n) = (\lambda z_1, \cdots, \lambda z_n).$$

The addition is associative and commutative because it is defined as addition of the components, which are complex numbers. Analogously the multiplication by a scalar is distributive.

We leave it to the reader to verify that all the axioms of a linear vector space are satisfied.

1.1 The C^n becomes a Banach space by introducing a norm $||\ \ ||$ satisfying: (1) $||z|| > 0$ if $z \neq 0$; (2) $||z^{(1)} + z^{(2)}|| \leq ||z^{(1)}|| + ||z^{(2)}||$; (3) $||\lambda z|| = (|\lambda|\ ||z||)$, where λ is a complex number; (4) the C^n is complete with respect to the norm; that is, if for a sequence $\{z^{(j)}\}$, $z^{(j)} \in C^n$ we have $||z^{(j)} - z^{(k)}||$ tending to zero as j and k tend to infinity, then there exists an element $z^{(0)} \in C^n$ such that

$$\lim_{j\to\infty} ||z^{(j)} - z^{(0)}|| = 0.$$

Examples of Norms. The euclidean norm: $||z||_e^2 = |z_1|^2 + \cdots + |z_n|^2$. The maximum norm: $||z||_m = \max\{|z_1|, \cdots, |z_n|\}$. Every norm induces a topology if one defines as neighborhoods of a point $z^{(0)}$ the point sets

$$\{z \mid ||z - z^{(0)}|| < \epsilon; \epsilon > 0\}.$$

It is easy to show (the reader may carry out the proof) that: For any norm $||\ \ ||$ there exist two numbers $\rho > 0$ and $\sigma > 0$ such that for any $z \in C^n$ we have

$$\rho||z||_m \leq ||z|| \leq \sigma||z||_m,$$

where $||\ \ ||_m$ is the maximum norm.

A consequence of this fact is: *In the C^n all the topologies generated by different norms are equivalent.*

1.2 An *open set* C^n is called a *region*, and an *open and connected set* is called a *domain.*

1.3 The C^n is *topologically* and as *additive group* isomorphic to the additive group of $2n$-tuples of real numbers R^{2n} if we associate $z \longrightarrow (x_1, \cdots, x_n, y_1, \cdots, y_n)$, where $z_j = x_j + iy_j$ and $||(x, y)|| = ||z||$.

2. LINEAR SUBSPACES

2.1 We say that the C^n is of "complex dimension n." A *linear subspace of complex dimension p* is a subset of the C^n that can be written in the form

$$\{z \mid z = \lambda_1 a_1 + \cdots + \lambda_p a_p;\ (\lambda_1, \cdots, \lambda_p) \in C^p\},$$

where $a_1, \cdots, a_p \in C^n$ and fixed.

2.2 A "translated linear subspace"

$$\{z \mid z = z^{(0)} + \lambda_1 a_1 + \cdots + \lambda_p a_p;\ (\lambda_1, \cdots, \lambda_p) \in C^p\},$$

$(z^{(0)}, a_1, \cdots, a_p \in C^n$ and fixed) we will call a *complex p-plane.*

Instead of being defined by such a parameter representation, a complex p-plane can also be given by $n - p$ linear equations.

It should be noted that while every complex p-plane C^n is also a real $2p$-plane in the associated R^{2n}, the converse is not true. There are real $2p$-planes in the R^{2n} that are not complex p-planes in the C^n.

The reader may verify that the real 2-plane

$$E = \{z \mid x_1 = 0,\ x_2 = 0\}$$

is not a complex 1-plane in the C^2.

3. SPECIAL DOMAINS

An arbitrary domain in C^n can be visualized directly only for $n = 1$ because already for C^2 the associated real space is of real dimension 4. One method of visualizing domains D in C^2 is to fix one of the four associated real variables, for instance, y_2, and to look at the intersections

$$D \cap \{z \mid y_2 = y_2^{(0)}\}$$

for various values of $y_2^{(0)}$.

Some domains with sufficient symmetry can also be represented by a point set in a real space R^2 or R^3.

Examples

3.1 *The hypersphere:* $\{z \mid |z_1|^2 + \cdots + |z_n|^2 < r\}$. (This is the "ball" of radius r in the euclidean norm.) For $n = 2$ it can be represented graphically as shown in Fig. 1.

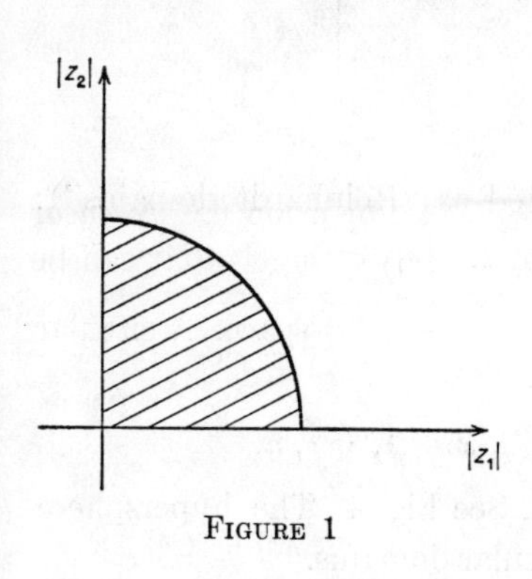

FIGURE 1

3.2 *The polycylinder:* $\{z \mid |z_1| < r_1, \cdots, |z_n| < r_n\}$. (For $r_1 = r_2 = \cdots = r_n = r$ this is the "ball" of radius r in the maximum norm.) The polycylinder (see Fig. 2) is the direct product of the n discs

$$\{z_1 \mid |z_1| < r_1\} \times \cdots \times \{z_n \mid |z_n| < r_n\}.$$

For complex dimension 1 both hypersphere and polycylinder coincide with the circle. For higher dimension they take with equal right the place of the circle, but they cannot even be mapped holomorphically onto each other. (This can be shown by means of invariants formed from Bergman's kernel function.)

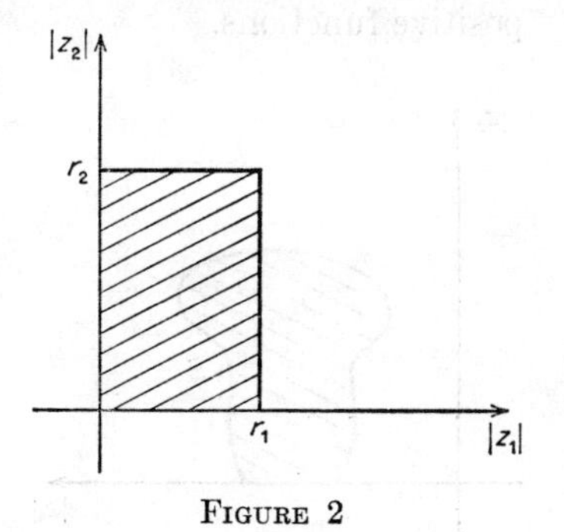

FIGURE 2

3.3 *Product domains:* $\{z \mid z_1 \in D_1, \cdots, z_n \in D_n\}$, where $D_1, \cdots, D_n$ are plane domains. See Fig. 3. The polycylinder is a product domain where the D_j are circles.

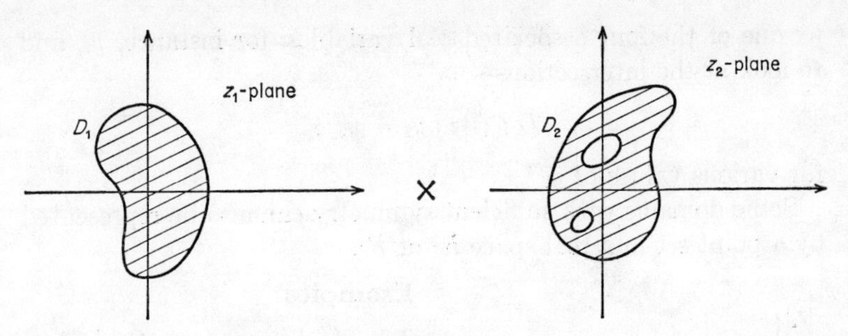

FIGURE 3

3.4 *Circular domains* (also denoted as "Reinhardt domains"):

$$\{z \mid (|z_1 - z_1^{(0)}|, \cdots, |z_n - z_n^{(0)}|) \in E\},$$

where E is a set in the $(|z_1 - z_1^{(0)}|, \cdots, |z_n - z_n^{(0)}|)$-space. A circular domain admits the automorphisms:

$$z_k^* - z_k^{(0)} = e^{i\theta_k}(z_k - z_k^{(0)}),\ k = 1, \cdots, n,$$

where $\theta_1, \cdots, \theta_n$ are arbitrarily real. See Fig. 4. The hypersphere and the polycylinder are special circular domains.

3.5 *Tube domains:* $\{z \mid x \in X, y \text{ arbitrary}\}$, $z_j = x_j + iy_j$, $x = (x_1, \cdots, x_n)$, $y = (y_1, \cdots, y_n)$, and X is a domain in the space of the real parts $(x_1, \cdots, x_n)$. See Fig. 5.

3.6 *Hartogs domains:* $\{(z, w) \mid z \in D, r(z) < |w - w^{(0)}| < R(z)\}$, where D is a domain in the C^n, $w \in C^1$, and $r(z)$ and $R(z)$ are positive functions.

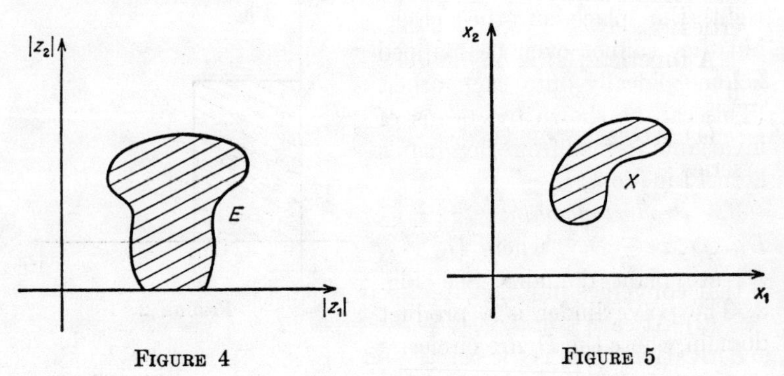

FIGURE 4 FIGURE 5

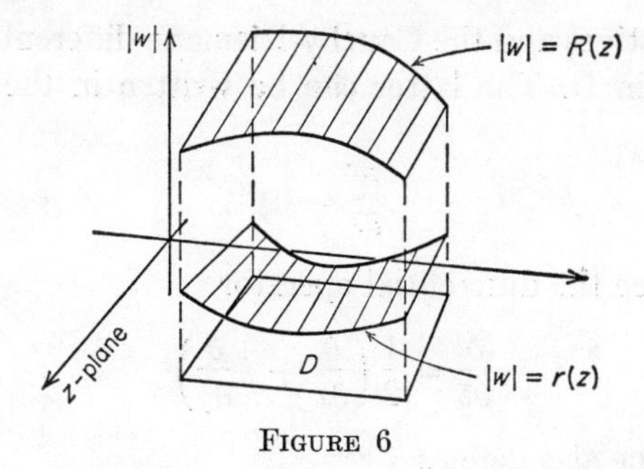

FIGURE 6

More generally, a Hartogs domain is a domain in (z, w)-space, $z \in C^n$, $w \in C^1$, that permits the following group of automorphisms: $z^* = z$, $w^* - w^{(0)} = e^{i\theta}(w - w^{(0)})$, θ arbitrary real. See Fig. 6.

4. HOLOMORPHIC FUNCTIONS

4.1 A function is the association of one and only one element in a certain "value set" to every element in an "argument set."

We will consider functions such that the values are complex (or real) and where the argument set is a domain in the C^n. p-tuples of such functions can be considered as one function with values in a C^p.

4.2 We remind the reader that the holomorphic functions of *one* complex variable can be characterized by four different properties:

A function $f(z)$ is holomorphic in a domain $D \subset C^1$ if and only if

(1) At each point $z^{(0)}$ of D it can be developed into a power series

$$f(z) = \sum_{\nu=0}^{\infty} a_\nu (z - z^{(0)})^\nu$$

that converges in a neighborhood of $z^{(0)}$.

(2) At each point $z^{(0)}$ of D the function $f(z)$ possesses a complex derivative. This is the case if and only if $f(z)$ possesses continuous

partial derivatives and the Cauchy-Riemann differential equations are satisfied in D. The latter can be written in the very simple form

$$\frac{\partial f}{\partial \bar{z}} = 0$$

if we introduce the differential operator

$$\frac{\partial}{\partial \bar{z}} = \frac{1}{2}\left(\frac{\partial}{\partial x} + i\frac{\partial}{\partial y}\right).$$

In addition one also defines

$$\frac{\partial}{\partial z} = \frac{1}{2}\left(\frac{\partial}{\partial x} - i\frac{\partial}{\partial y}\right).$$

(3) $w = f(z)$ maps the neighborhood of any point $z^{(0)} \in D$ at which $f'(z^{(0)}) \neq 0$ *conformally*. (That means: given two curves through z_0 and the angle between their tangents, then the angle between the tangents of the image curves in the w-plane is the same, in magnitude and direction.)

(4) $f(z)$ is holomorphic in D if and only if $f(z)$ is continuous in D and the integral $\int_{z^{(0)}}^{z} f(\zeta)\, d\zeta$; $z^{(0)}, z \in D$, is locally independent of the path of integration. (Cauchy's theorem and Morera's theorem.)

Each of these properties can be generalized to several variables and defines a class of functions. The question arises: are these classes of functions identical as they are for one variable?

4.3 DEFINITION: *A complex-valued function $f(z)$ defined in a domain $D \subset C^n$ is called* holomorphic in D in the sense of Weierstrass *if it can be developed at each point $z^{(0)}$ of D into a multiple power series*

$$f(z) = \sum_{\nu_1, \ldots, \nu_n = 0}^{\infty} a_{\nu_1, \ldots, \nu_n}(z_1 - z_1^{(0)})^{\nu_1}, \cdots, (z_n - z_n^{(0)})^{\nu_n}$$

that converges uniformly in a neighborhood of $z^{(0)}$.

4.4 DEFINITION: *A complex-valued function $f(z)$ defined in a domain $D \subset C^n$ is called* holomorphic in D in the sense of Cauchy-

Riemann *if and only if at every point* $z^{(0)}$ *in* D *the partial derivatives exist and the Cauchy-Riemann differential equations*

$$\frac{\partial f}{\partial \bar{z}_j} = 0, \qquad j = 1, \cdots, n$$

are satisfied, where

$$\frac{\partial}{\partial \bar{z}_j} = \frac{1}{2}\left(\frac{\partial}{\partial x_j} + i\frac{\partial}{\partial y_j}\right).$$

This means that if we replace in $f(z)$ any $n - 1$ variables by constants, then the remaining function is a holomorphic function (in the sense of one variable) of z_j.

We also can say: $f(z)$ is a holomorphic function of each variable separately.

4.5 *"W-holomorphic" implies "CR-holomorphic."*

Indeed, if $f(z)$ is W-holomorphic, then

$$f(z) = \sum_0^\infty a_{\mu_1, \cdots, \mu_n}(z_1 - z_1^{(0)})^{\mu_1}, \cdots, (z_n - z_n^{(0)})^{\mu_n}.$$

The reader may show: If a multiple power series converges for a point $z_1^{(1)}, \cdots, z_n^{(1)}$, then it converges in any polycylinder

$$\{z \mid |z_1 - z_1^{(1)}| \le r_1, \cdots, |z_n - z_n^{(1)}| \le r_n\}$$

with

$$r_1 < |z_1^{(1)} - z_1^{(0)}|, \cdots, r_n < |z_n^{(1)} - z_n^{(0)}| \quad \textit{uniformly.}$$

Hence the power series above converges uniformly in a neighborhood of $z^{(0)}$. Hence, by a well-known theorem on uniform convergence, we may differentiate the series term by term.

Now

$$\frac{\partial}{\partial \bar{z}_j}(z_1 - z_1^{(0)})^{\mu_1}, \cdots, (z_n - z_n^{(0)})^{\mu_n} \equiv 0.$$

Hence $f(z)$ is CR-holomorphic.

4.6 For $n = 1$ one proves that a CR-holomorphic function can be developed in a power series as follows:

$$f(z) = \frac{1}{2\pi i}\int_C \frac{f(\zeta)}{\zeta - z}\,d\zeta = \frac{1}{2\pi i}\int_C \frac{f(\zeta)}{(\zeta - z_0) - (z - z_0)}\,d\zeta$$

where C is a suitable circle around z_0. Then one develops

$$\frac{1}{(\zeta - z_0) - (z - z_0)} = \frac{1}{(\zeta - z_0)}\left(\frac{1}{1 - \dfrac{z - z_0}{\zeta - z_0}}\right) = \sum_{\nu=0}^{\infty} \frac{(z - z_0)^\nu}{(\zeta - z_0)^{\nu+1}},$$

exchanges summation and integration, and has the Taylor development of $f(z)$. The exchange of integration is permissible because the geometric series converges uniformly and $f(\zeta)$ is continuous. The latter fact is a consequence of the existence of the complex derivative f'.

It is not difficult to generalize Cauchy's integral formula for CR functions of several variables for product domains.

Let $f(z)$ be a CR-holomorphic in a neighborhood of the product domain $D = D_1 \times \cdots \times D_n$. Then

$$f(z) = \frac{1}{(2\pi i)^n} \int_{\partial D_1 \times \cdots \times \partial D_n} \frac{f(\zeta)}{(\zeta_1 - z_1), \cdots, (\zeta_n - z_n)}\, d\zeta_1, \cdots, d\zeta_n.$$

Here we integrate over the direct product of the boundaries $\partial D_1, \cdots, \partial D_n$. (Please note that this product has (real) dimension n, while the topological boundary of D has dimension $2n - 1$.)

We can develop each of the factors $1/(\zeta_j - z_j)$ in a geometric series (analogously to one variable). However, we cannot conclude that $f(\zeta)$ is continuous.

The function $f(x, y) = xy/(x^4 + y^4)$, $f(0, 0) = 0$, is an example of a function of two variables such that $\partial f/\partial x$ and $\partial f/\partial y$ exist everywhere. The function is continuous in each variable separately, but not in both. It is not even bounded!

This prevents us from exchanging integration and summation as we did for one variable. Hence this proof works only under the additional assumption that $f(z)$ is continuous in D (or at least locally bounded).

To prove that this assumption is superfluous, or, what is equivalent, to prove that *if $f(z)$ is CR-holomorphic, then it is continuous*, was one of the first outstanding problems of the theory. It was solved by F. Hartogs [26]. His proof uses theorems of measure theory. All later proofs (that have not been found faulty) follow his basic line of reasoning. The proof is a bit too involved to be reproduced in this limited space. A good version of it, and rather independently readable, can be found in Carathéodory [17, vol. 2]. Thus we have:

4.7 THEOREM: *A function is W-holomorphic if and only if it is CR-holomorphic.*

Since the two notions of holomorphy are equivalent we will in the following speak only of *holomorphic functions.* As a corollary of Theorem 4.7 we have: *A holomorphic function is continuous.*

4.8 While a holomorphic function of one variable maps *conformally,* this is not true for holomorphic functions of more than one variable.

The reader may prove: Let D be a domain in C^n. Let $f_1(z), \cdots, f_n(z)$ be n holomorphic functions in D. Then

$$z_1^* = f_1(z), \cdots, z_n^* = f_n(z)$$

maps a neighborhood of any point $z^{(0)}$ one-to-one onto a neighborhood of $z^{(0)*} = (f_1(z^{(0)}), \cdots, f_n(z^{(0)}))$ provided that the Jacobian

$$\begin{vmatrix} \dfrac{\partial f_1}{\partial z_1} & \cdots & \dfrac{\partial f_1}{\partial z_n} \\ \cdots & \cdots & \cdots \\ \dfrac{\partial f_n}{\partial z_1} & \cdots & \dfrac{\partial f_n}{\partial z_n} \end{vmatrix}$$

does not vanish at $z^{(0)}$.

The following is an example that such a mapping need not be conformal. The mapping

$$z_1^* = z_1 + z_2, \qquad z_2^* = z_2$$

has Jacobian $\equiv 1$. It maps the two curves

$$\mathfrak{C}_1 = \{z \mid z_1 = t_1, z_2 = t_1, -\infty < t_1 < \infty\}$$

and

$$\mathfrak{C}_2 = \{z \mid z_1 = t_2, z_2 = 0, -\infty < t_2 < \infty\}$$

which intersect at an angle of 45° into the two curves

$$\mathfrak{C}_1^* = \{z^* \mid z_1^* = 2t_1, z_2^* = t_1, -\infty < t_1 < \infty\}$$

and

$$\mathfrak{C}_2^* = \{z^* \mid z_1^* = t_2, z_2^* = 0, -\infty < t_2 < \infty\},$$

which no longer make an angle of 45°.

However, two curves lying on a complex 1-plane are mapped conformally. For this reason the mappings by n-tuples of holomorphic functions of n variables are often called "pseudo-conformal." We prefer to use the term "holomorphic mapping."

4.9 The fourth property that characterizes holomorphic functions of one variable can be generalized; however, instead of a curve we have to integrate over a closed $(2n - 1)$-dimensional surface, and instead of one integral we get n integrals. This generalization is of little interest.

4.10 The real and imaginary parts of holomorphic functions of one variable are *harmonic functions*, which means they satisfy the Laplacian differential equation

$$4 \frac{\partial^2 h}{\partial z \, \partial \bar{z}} = \frac{\partial^2 h}{\partial x^2} + \frac{\partial^2 h}{\partial y^2} = 0.$$

And conversely, for every harmonic function $\phi(x, y)$ there exists a conjugate harmonic function $\psi(x, y)$ such that $\phi(x, y) + i\psi(x, y)$ is locally holomorphic. (In the large, $\phi + i\psi$ need not be single-valued. Hence we say "locally holomorphic.")

The reader may show (as an immediate consequence of the Cauchy-Riemann differential equations) that the real (and imaginary) part of a holomorphic function of n variables satisfies the differential equations:

$$\frac{\partial^2 h}{\partial z_\mu \, \partial \bar{z}_\nu} = 0 \qquad \text{for } \mu, \nu = 1, \cdots, n.$$

This implies that h is harmonic in each pair of variables x_j, y_j $\left(\text{because } \dfrac{\partial^2 h}{\partial z_\mu \, \partial \bar{z}_\mu} = 0\right)$, and that h is a harmonic function of the $2n$ real variables $x_1, y_1, \cdots, x_n, y_n$ because

$$\Delta^{(n)} h = \frac{\partial^2 h}{\partial x_1^2} + \frac{\partial^2 h}{\partial y_1^2} + \cdots + \frac{\partial^2 h}{\partial x_n^2} + \frac{\partial^2 h}{\partial y_n^2} = 0.$$

The converse is not true, of course. A harmonic function, in general, is not the real part of a holomorphic function.

The equations

$$\frac{\partial^2 h}{\partial z_\mu \, \partial \bar{z}_\nu} = 0 \qquad \text{for } \mu, \nu = 1, \cdots, n$$

are not only necessary that h is the real part of a locally holomorphic function, but also sufficient. (This the reader may show.) The functions satisfying the equations are called *pluriharmonic*.

The system is overdetermined for solving boundary-value problems (which are uniquely solvable in the larger class of harmonic functions, provided that the boundary and the prescribed values are "sufficiently smooth").

The question of what boundary-value problems *can* be solved by the pluriharmonic functions (as well as by slightly larger classes) has been studied by S. Bergman in several papers (comp. [6]) and in Bremermann [14].

5. POWER SERIES, SIMULTANEOUS HOLOMORPHIC CONTINUATION

5.1 We have already stated: if a multiple power series

$$\sum_{0}^{\infty} a_{\nu_1 \cdots \nu_n}(z_1 - z_1^{(0)})^{\nu_1} \cdots (z_n - z_n)^{\nu_n}$$

converges for a point $z^{(1)}$, then it converges uniformly (in all variables) and absolutely in every polycylinder

$$\{z \mid |z_1 - z_1^{(0)}| < r, \cdots, \quad |z_n - z_n^{(0)}| < r_n\},$$

with $$r_1 < |z_1^{(1)} - z_1^{(0)}|, \cdots, \quad r_n < |z_n^{(1)} - z_n^{(0)}|.$$

This does not imply that the maximal domain of convergence of a multiple power series is a polycylinder. A closer investigation gives (comp. Bremermann [14] and Bers [7]):

THEOREM: *The maximal region of uniform convergence of a multiple power series is always a circular domain*

$$D = \{z \mid (|z_1 - z_1^{(0)}|, \cdots, |z_n - z_n^{(0)}|) \in E\}$$

such that the image of E under the mapping $s_1 = \log |z_1 - z_1^{(0)}|, \cdots,$ $s_n = \log |z_n - z_n^{(0)}|$ is a convex domain in the (real) $s_1, \cdots, s_n$-space.

5.2 The radii $r_1, \cdots, r_n$ of any maximal polycylinder contained in D are called *associated radii of convergence*. $n - 1$ of the radii can be chosen independently; then the remaining radius is determined. Each radius $r_j(r_1, \cdots, \hat{r}_j, \cdots, r_n)$ is a monotonically decreasing function of the remaining radii $r_1, \cdots, \hat{r}_j, \cdots, r_n$ ($\hat{r}_j$ indicates that the variable r_j is missing).

5.3 *The result* 5.1 (*but not* 5.2) *remains true for multiple Laurent series:*

$$\sum_{-\infty}^{+\infty} a_{\nu_1 \cdots \nu_n}(z_1 - z_1^{(0)})^{\nu_1} \cdots (z_n - z_n^{(0)})^{\nu_n}.$$

(The sum is to be understood such that $\nu_1, \cdots, \nu_n$ range independently of each other from $-\infty$ to $+\infty$.)

5.4 THEOREM: *Given any circular domain*

$$D = \{z \mid (|z_1 - z_1^{(0)}|, \cdots, |z_n - z_n^{(0)}|) \in E\},$$

where E is such that D is a domain but otherwise arbitrary. Then any function that is holomorphic in D can be developed into a multiple Laurent series that converges uniformly not only in D but in (every closed subset of) the circular domain

$$D^* = \{z \mid (|z_1 - z_1^{(0)}|, \cdots, |z_n - z_n^{(0)}|) \in E^*\}$$

where E^* is such that its logarithmic image is the convex envelope of the logarithmic image of E.

This Laurent development is unique. (Comp. Bremermann [15] and Bers [7].)

5.5 Theorem 5.4 has a remarkable consequence, unknown for one variable: The Laurent development of a function holomorphic in D converges still in D^*. Because of the uniform convergence we may exchange differentiation and summation. Hence the series converges to a holomorphic function in D^*.

DEFINITION: *Given a domain D and a function $f(z)$ holomorphic in D and a domain D^* containing D and a function $g(z)$ holomorphic in D^* such that $f(z) \equiv g(z)$ in D. Then $g(z)$ is called a* holomorphic continuation *of $f(z)$ into D^*.*

Holomorphic continuation is *unique* in the sense: *If $g_1(z)$ and $g_2(z)$ are holomorphic continuations of a function $f(z)$ from D into D^* (and D is not empty) then $g_1(z) \equiv g_2(z)$ in D^*.*

The reader may derive this as an immediate consequence of

the uniqueness of holomorphic continuation in one complex variable.

Thus Theorem 5.4 has the following corollary, which has no analogue for one complex variable:

COROLLARY: *If $f(z)$ is holomorphic in a circular domain*

$$D = \{(z_1, \cdots, z_n) \mid (|z_1 - z_1^{(0)}|, \cdots, |z_n - z_n^{(0)}|) \in E\}$$

then $f(z)$ possesses a holomorphic continuation into the circular domain

$$D^* = \{(z_1, \cdots, z_n) \mid (|z_1 - z_1^{(0)}|, \cdots, |z_n - z_n^{(0)}|) \in E^*\},$$

where E^* is such that its logarithmic image is the convex envelope of the logarithmic image of E.

For $n = 1$ we have $E = E^*$; for $n > 1$ there exist sets E such that $E \neq E^*$.

The function f in the corollary is an arbitrary function. Hence we have: All functions that are holomorphic in D possess a holomorphic continuation into D^*. This phenomenon is called *simultaneous holomorphic continuation.* We will further explore it in Sec. 6.

Example. See Fig. 7.

$$E = \{|z_1| < 1, |z_2| < r_2\} \cup \{r_1 < |z_1| < 1, |z_2| < 1\}.$$

$$0 < r_1 < 1, \qquad 0 < r_2 < 1.$$

Then

$$E^* = \{|z_1| < 1, \qquad |z_2| < 1\}.$$

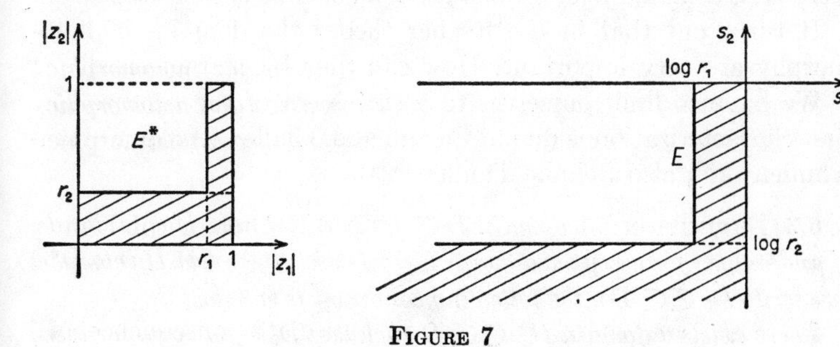

FIGURE 7

6. DOMAINS OF HOLOMORPHY

6.1 In the preceding example we have a domain D such that all functions holomorphic in D can be continued into a larger domain D^*.

There exist domains that do not have this property. For instance the function $\sum_{n=0}^{\infty} z^{n!}$ converges in the unit disc to a function that cannot be continued beyond the unit circle. (Comp. Titchmarsh [49].) It can be shown generally (comp. Behnke-Sommer [1, p. 566]): *For every domain $D \subset C^1$ there exists a function that is holomorphic in D and that cannot be continued holomorphically into any boundary point of D.* (In this case the boundary of D is the *natural boundary* for the function.) This result is remarkable because we do not impose any restriction on the boundary of D. As the reader knows from examples in set theory, the boundary of a connected open point set which is otherwise arbitrary can be quite complicated.

6.2 DEFINITION: *A domain $D \subset C^n$ with the property that there exists a function that is holomorphic in D and that cannot be continued locally into any boundary point of D is called* a domain of holomorphy.

Thus we can express the previous statements as follows: (1) *Every domain $D \subset C^1$ is a domain of holomorphy.* (2) *For $n > 1$ there exist domains $D \subset C^n$ that are not domains of holomorphy.*

It turns out that in the further theory the domains of holomorphy are very important. How can they be characterized?

We have to limit ourselves to a statement of the results. The first characterization is due to Cartan and Thullen (comp. Cartan-Thullen [20], also Behnke-Thullen [2]).

6.3 DEFINITION: *A domain $D \subset C^n$ is called* holomorph-convex *if and only if for every subdomain D_0 of D which is relatively compact (we write $D_0 \subset\subset D$), the following condition is satisfied:*

There exists a domain $D^* \subset\subset D$ such that $D_0 \subset D^*$ and for every *point $z_0 \in D - D^*$ there exists a function $f(z)$ that is holomorphic in D and satisfies the condition*

$$\sup_{z \in D_0} |f(z)| < |f(z_0)|.$$

Since we deal with $2n$-dimensional domains, it is difficult to visualize this situation (which for $n = 1$ is trivial and is satisfied by any domain D).

We obtain an analogue if we replace D by a domain in R^2 and the class of moduli of functions $f(z)$ that are holomorphic in D by *linear* functions of two real variables $l(x)$. The condition then can be interpreted as shown in Fig. 8. For every $D_0 \subset\subset D$ a D^* and $l(x)$ can be found such that $l(x) = l(x_0)$ *does not intersect* D_0 (because $l(x_0) > \sup_{x \in D_0} l(x)$). This implies that D *is convex*, and conversely if D is convex, then the condition is satisfied.

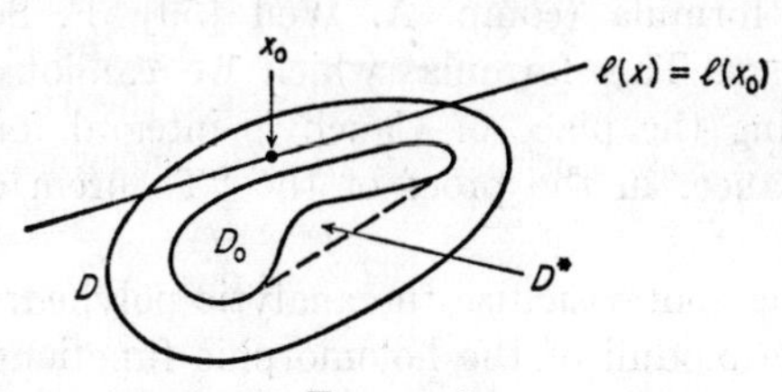

FIGURE 8

This analogy with convex domains is indicated in the term *holomorph-convex*. The following theorem was first proved by Cartan and Thullen [20] (comp. also Behnke-Thullen [2], Bremermann [15]).

6.4 THEOREM OF CARTAN-THULLEN: *A domain $D \subset C^n$ is a domain of holomorphy if and only if D is holomorph-convex.*

As a consequence of the theorem of Cartan-Thullen the reader may prove (by means of the theorem of Heine-Borel) the following result:

6.5 DEFINITION: *A region $P \subset C^n$ is called an* analytic polyhedron *if and only if it is of the following form:*

$$P = \{(z_1, \cdots, z_n) \mid |f_1(z)| < 1, \cdots, |f_k(z)| < 1\},$$

where $f_1(z), \cdots, f_k(z)$ *are holomorphic functions in some domain* $D \subset C^n$ *such that* $P \subset\subset D$.

It can be shown that *the number* k *of the defining function* $f_1, \cdots, f_k$ *is larger than or equal to* n, otherwise it is not possible that $P \subset\subset D$.

THEOREM: *Every domain of holomorphy can be approximated from the interior by analytic polyhedra* $P_\nu \subset\subset D$ *such that the defining functions are holomorphic in* D, *and*

$$P_\nu \subset P_{\nu+1} \subset\subset D \quad \textit{and} \quad \lim_{\nu\to\infty} P_\nu = D.$$

Here "$\lim_{\nu\to\infty} P_\nu = D$" has the following meaning: for every subdomain $G \subset\subset D$ there exists a ν_0 such that for $\nu > \nu_0$, $G \subset P_\nu$.

6.6 Analytic polyhedra are of special interest. The Weil-Bergman integral formula (comp. A. Weil [50], F. Sommer [45]) applies to them. This formula, which we cannot give here in detail, is taking the place of Cauchy's integral formula and is used, for instance, in the proof of the "Theorem of Oka" (see below).

Again, one cannot visualize the analytic polyhedra directly. If we replace the moduli of the holomorphic functions f_j by linear functions l_j, then we obtain *convex polyhedra.* The limits of convex polyhedra are convex domains, and conversely, every convex domain can be approximated by convex polyhedra.

This analogy, however, should not tempt us to think of the holomorph-convex domains as slightly generalized convex domains. We should keep in mind that (according to the theorem of Cartan-Thullen and to 6.2) *every domain in the complex plane is holomorph-convex.* Also, while every domain of holomorphy can be approximated by analytic polyhedra from the *interior*, there are examples that show that an analogous approximation from the *exterior* is not always possible (Behnke-Thullen [2]). This peculiarity does not occur for convex domains.

6.7 THEOREM OF BEHNKE-STEIN: *Let* D_ν *be a sequence of domains of holomorphy such that* $D_\nu \subset D_{\nu+1}$. *Then* $\lim_{\nu\to\infty} D_\nu$ *is a domain of holomorphy.*

(The limit of the sequence may be the whole space C^n.) This theorem, plausible as it is, again is not completely trivial. In the proof one makes use of results of the type of Theorem 7.3.

6.8 The following result, however, is not difficult and the reader may easily obtain it from the definition.

THEOREM: *The largest open set contained in the intersection of a finite or infinite number of holomorph-convex domains is holomorph-convex.*

Because of the theorem of Cartan-Thullen the same result holds for domains of holomorphy.

6.9 The characterization of the domains of holomorphy as holomorph-convex domains is, in some sense, a *global* characterization. Given a domain, it is not easy to decide whether the criterion is satisfied.

There is a different characterization which is local and purely geometric. It is relatively easy to prove that this property, which is called "pseudo-convex," is necessary. This was first shown by E. E. Levi. To prove or disprove the converse remained a major problem till it was solved by Oka in 1942 (for $n = 2$) and by Oka, Norguet, and Bremermann (for arbitrary n) in 1954.

We will state the notion "pseudo-convex" in a convenient modern form. There are many equivalent definitions, some of them applicable only to a restricted class of domains.

6.10 DEFINITION: *Let D be a domain in C^n. Let*

$$d_D(z) = \sup r \ni \{z' \mid ||z' - z|| < r\} \subset D,$$

where

$$||z' - z||^2 = |z_1' - z_1|^2 + \cdots + |z_n' - z_n|^2.$$

In other words, $d_D(z)$ is the radius of the largest ball with center at z that is contained in D; or again in other words: $d_D(z)$ is the distance of the point z from the boundary of z, measured in the euclidean norm.

The reader may easily prove, directly from the definition:

If D has at least one finite boundary point, then $d_D(z)$ is finite and continuous throughout D. If $D = C^n$, then $d_D(z) \equiv \infty$.

We will, in what follows, use this function $d_D(z)$ to define the "pseudo-convex domains" after having defined the "plurisubharmonic functions."

6.11 Definition: *A real-valued function $V(z)$ defined in a domain $D \subset C^1$ is called* subharmonic *if and only if it satisfies the following three conditions:*

(a) $-\infty \leq V(z) < \infty$.

(b) *$V(z)$ is upper semicontinuous.*

(c) *If $G \subset\subset D$ and $h(z)$ is a harmonic function in G, continuous in $\overline{G}$, then if the inequality $h(z) \geq V(z)$ is satisfied on the boundary ∂G of G, then it is satisfied in the interior of G.*

The condition (c) can be replaced by the following:

(c′) $\dfrac{\partial^2 V}{\partial x^2} + \dfrac{\partial^2 V}{\partial y^2} \geq 0$ *in D* (the derivatives taken in the sense of Schwartz' distributions).

6.12 The subharmonic functions were first investigated by F. Hartogs (for a bibliography see Bremermann [8]). He encountered them in the study of series of the form $\sum_{n=0}^{\infty} a_n(z)w^n$ ("Hartogs series"), where the "coefficients" $a_n(z)$ are holomorphic functions in a domain D of the z-plane. It turns out that the region of uniform convergence (in both w and z) is a "Hartogs domain" $\{(z, w) \mid z \in D, |w| < R(z)\}$. The function $R(z)$ is not arbitrary, but such that $-\log R(z)$ is a subharmonic function in D.

If one replaces D by a domain in C^n, then one gets the same result, except that the function $\log R(z)$ is not subharmonic but "plurisubharmonic," which we define in the following. The plurisubharmonic functions, however, have considerable importance beyond this point.

6.13 Definition: *A real-valued function $V(z)$, defined in a domain $D \subset C^n$, is called* plurisubharmonic *if and only if it satisfies the following conditions:*

(a) $-\infty \leq V(z) < \infty$.

(b) *$V(z)$ is upper semicontinuous.*

(c) *The restriction of $V(z)$ to any complex 1-plane $L = \{z \mid z = z_0 + \lambda a\}$ is a subharmonic function (of λ) in the intersection $L \cap D$.*

The condition (c) can be replaced by the following:

(c′) $\sum_{\mu,\nu=1}^{n} \frac{\partial^2 V}{\partial z_\mu \, \partial \bar{z}_\nu} a_\mu \bar{a}_\nu \geq 0$ *for all n-tuples of complex numbers* $(a_1, \cdots, a_n)$ (the derivatives taken in the sense of Schwartz' distributions).

For $n = 1$ the plurisubharmonic functions and the subharmonic functions are identical. The reader will have no difficulty in deriving condition (c′) above from condition (c′) of 6.11.

$$\text{Note that } \frac{\partial^2}{\partial z \, \partial \bar{z}} = \frac{1}{4}\left(\frac{\partial^2}{\partial x^2} + \frac{\partial^2}{\partial y^2}\right).$$

6.14 The plurisubharmonic functions have certain *algebraic properties* in common with the convex functions of n real variables.

Convex functions may be defined quite in analogy to the subharmonic and plurisubharmonic functions. Conditions (a) and (b) become superfluous, and in 6.11(c) one replaces the *harmonic majorants* by *linear majorants.* The differential conditions also apply directly to the convex function, if we replace $\frac{\partial^2}{\partial z_\mu \, \partial \bar{z}_\nu}$ by $\frac{\partial^2}{\partial x_\mu \, \partial x_\nu}$. Both classes of functions form a *convex cone.* A convex cone is a set of functions S satisfying the following conditions:

I. If $\phi_1, \phi_2 \in S$, then $\phi_1 + \phi_2 \in S$.

II. If $c \geq 0$ and $\phi \in S$, then $c\phi \in S$.

The reader will find a relatively self-contained treatment of the similarities between the convex function and the plurisubharmonic functions in Bremermann [11].

6.15 Theorem: *Let G be a domain in the real space R^n. Let $d_G(x)$ denote the distance of the point x from the boundary of G ($d_G(x)$ is defined as in* 6.10).

Then G is convex if and only if $-\log d_G(x)$ is a convex function in G (Bremermann [11]).

Thus one could define the convex domains by requiring that $\log d_G(x)$ is convex. There is no sense in doing this because the standard definition is simpler. We will, however, use $-\log d$ and the plurisubharmonic function to define the pseudo-convex domains. The analogue with the convex functions should provide some insight.

6.16 DEFINITION: *Let $D \subset C^n$. Then D is called* pseudo-convex *if and only if $-\log d_D(z)$ is plurisubharmonic in D.*

The pseudo-convex domains have numerous interesting properties. (The reader will find a fairly easily accessible treatment in Bremermann [11].)

The following theorem is very deep and powerful. One direction ("domains of holomorphy are pseudo-convex") is fairly easy to prove. The converse constituted a major problem for about thirty years. Its proof uses all the notions mentioned so far plus several other powerful theorems stated in the following. In some sense, it is a transition from a *local* property of a domain to a *global* property. Results about transitions from local to global properties occur also in other branches of mathematics. As a rule, they are difficult to prove.

6.17 THEOREM OF OKA: *A domain $D \subset C^n$ is a domain of holomorphy if and only if D is pseudo-convex* (Oka [37] [40]; Norguet [31]; Bremermann [10]).

6.18 Pseudo-convex domains (and hence domains of holomorphy) have peculiar properties, in particular: I. *The exterior of a bounded set cannot be a pseudo-convex domain except for $n = 1$.* II. *The boundary of a pseudo-convex domain (and hence of a domain of holomorphy) is a connected set (except for $n = 1$).*

COROLLARY: *A holomorphic function cannot have isolated points as singularities except for $n = 1$.*

7. RUNGE DOMAINS

7.1 Given a domain D in the complex z-plane. Under what conditions is it possible to approximate any function $f(z)$ holomorphic in D by polynomials $p(z)$ in the uniform norm? (We say we can approximate a given function $f(z)$ in the "uniform norm" by polynomials if and only if the following condition holds: For every $D_0 \subset\subset D$ and for every $\epsilon > 0$ there exists a polynomial $p(z)$ such that $|f(z) - p(z)| < \epsilon$ for $z \in D_0$.)

The answer to the question is given by the well-known theorem of Runge:

Let D be a domain of the z-plane. Let D be simply connected. Then every function holomorphic in D can be approximated by polynomials (in the uniform norm). And conversely, if all functions can be so approximated, then D is simply connected.

7.2 It has been demonstrated that Runge's theorem in the above form becomes false in both directions if we replace C^1 by C^n. There are domains with the topological structure of the cell such that certain functions cannot be approximated by polynomials, and conversely there exist domains of an almost arbitrary topological complication, and still all functions can be approximated.

To get a simple characterization one must require that D is a domain of holomorphy.

DEFINITION: *$D \subset C^n$ is called a* Runge domain *if and only if D is a domain of holomorphy and all its holomorphic functions can be approximated by polynomials (in the uniform norm).*

7.3 Like the domains of holomorphy, the Runge domains can be characterized in two different ways:

THEOREM (due to A. Weil [51]): *D is a Runge domain if and only if D is "polynomially convex."*

Here polynomially convex means the following: We take the definition of "holomorph-convex" and require that the functions f, occurring there in the "majorization condition," $\sup_{z \in D_0} |f(z)| < |f(z_0)|$, are polynomials.

7.4 THEOREM: *A domain $D \subset C^n$ is a Runge domain if and only if D is of the form*

$$D = \lim_{\nu \to \infty} D_\nu, \qquad D_\nu \subset D_{\nu+1} \subset D,$$

where $D_\nu = \{z \mid V_\nu(z) < 0\}$, *where* $V_\nu(z)$ *is a function that is plurisubharmonic in the whole* C^n (Bremermann [12]).

The connection with the pseudo-convex domains becomes clear if we observe: A domain D is pseudo-convex if and only if $D = \lim_{\nu \to \infty} D_\nu$, $D_\nu \subset D_{\nu+1} \subset D$, where $D_\nu = \{z \mid V_\nu(z) < 0\}$, $V_\nu(z)$ plurisubharmonic in D (instead of plurisubharmonic in C^n). Therefore domains satisfying the condition of 7.6 have also been called "pseudo-convex relative to the C^n." (Comp. Bremermann [12]; for generalizations to "complex spaces" see Grauert [23].)

7.5 If D is a Runge domain then every connected component of the intersection $D \cap L$ of D with any complex 1-plane $L = \{z \mid z = z_0 + \lambda a\}$ is *simply connected.* This is an easy consequence of Theorem 7.4. Up to the present day *it is an open problem whether the condition that the components of* $D \cap L$ *are simply connected is sufficient for* D *to be a Runge domain, even under the additional assumption that* D *is a domain of holomorphy.*

7.6 A sphere is a Runge domain. It can be shown that the union of two nonintersecting spheres is a Runge *region* (defined like a Runge domain except that it need not be connected. A holomorphic function in a "region" consists of a holomorphic function in each connected component; the functions in the various components are *independent* of each other). E. Kallin [28a] has shown that the union of any three nonintersecting spheres is a Runge region. However, for convex domains, instead of spheres there are counterexamples.

8. ANALYTIC SURFACES AND ANALYTIC SETS

We have not yet spoken about the zeros of a holomorphic function.

8.1 THEOREM: *A holomorphic function of more than one variable never has isolated points as zeros.*

In other words: If $f(z)$ is holomorphic in a neighborhood of z_0 and vanishes at z_0, then it is not possible that $f(z) \neq 0$ for all $z \neq z_0$ (in a neighborhood).

This is a consequence of a property of pseudo-convex domains:

$(f(z))^{-1}$ is holomorphic if $f(z) \neq 0$. The function $(f(z))^{-1}$ would have an isolated singularity. This is impossible (comp. 6.18).

8.2 Similarly it can be shown:

Let $f(z)$ be holomorphic in D. Then the point set $\{z \mid f(z) = 0\}$ has points arbitrarily close to the boundary of D.

8.3 THEOREM: *In the neighborhood of a zero, $(z_1^{(0)}, \cdots, z_n^{(0)})$ a function $f(z_1, \cdots, z_n)$ can be written, provided that $f(z_1, z_2^{(0)}, \cdots, z_n^{(0)} \not\equiv 0$:*

$$f(z_1, \cdots, z_n) = ((z_1 - z_1^{(0)})^m A_m(z_2, \cdots, z_n) + \cdots + A_0(z_2, \cdots, z_n))\Omega(z_1, \cdots, z_n)$$

where $\Omega(z_1^{(0)}, \cdots, z_n^{(0)}) \neq 0$, and $A_m(z_2, \cdots, z_n), \cdots, A_0(z_2, \cdots, z_n)$ are holomorphic functions of $z_2, \cdots, z_n$, $A_j(z_2^{(0)}, \cdots, z_n^{(0)}) = 0$.

The expression $(z_1 - z_1^{(0)})^m A_m(z_2, \cdots, z_n) + \cdots + A_0(z_2, \cdots, z_n)$ is called a "pseudo-polynomial." The theorem is known as the "Weierstrass preparation theorem" (comp. Bochner-Martin [3, p. 188]).

8.4 *The totality of the zeros of a holomorphic function is called "analytic surface"*—in general it is a $(2n - 2)$-dimensional surface with singularities.

A set S with the property that for every point $z_0 \in S$ there exists a neighborhood N such that in $S \cap N$ the set S consists of the simultaneous zeros of a finite number of functions holomorphic in N is called *analytic set.*

More information about analytic sets can be found in Remmert-Stein [43].

9. THE COUSIN PROBLEMS

9.1 Given a finite number of points $z^{(1)}, \cdots, z^{(m)}$ in a domain D in the complex plane and orders $\nu_1, \cdots, \nu_m$, then the product $(z - z^{(1)})^{\nu_1}, \cdots, (z - z^{(m)})^{\nu_m}$ vanishes at the points $z^{(1)}$ to $z^{(m)}$ and only at these in the prescribed orders. If instead of finitely many points there are infinitely many that do not have a limit point within D, then there still exists a function $F(z)$ vanishing in the prescribed orders at the prescribed points and nonzero otherwise. (This is a classical result. Compare Behnke-Sommer [1].)

9.2 One also can formulate this result as follows (the reader may show the equivalence):

Given for each point of $z_0 \in D$ *a neighborhood* $N(z^{(0)})$ *and a "local function"* $f_{z_0}(z)$ *holomorphic in* $N(z^{(0)})$. *Given also that the local functions satisfy the following "compatibility condition": If* $N(z^{(j)}) \cap N(z^{(k)}) \neq 0$, *then* $f_{z_j}(z)/f_{z_k}(z)$ *is holomorphic and nonzero.*

Then there exists a function $F(z)$ *holomorphic in* D *such that* $F(z)/f_{z_0}(z)$ *is holomorphic and nonzero in* $N(z_0)$ *for all points* z_0.

The *multiplicative Cousin problem* asks whether the same statement is true for more than one variable. For $n > 1$ the situation is, of course, much more delicate since the zeros are not isolated points but full surfaces.

The answer is: The multiplicative Cousin problem *cannot generally be solved if* D *is an arbitrary domain* (comp. Behnke-Thullen [2, p. 68]). It *always has a solution* if D is a domain of holomorphy and if there is *a continuous solution*, that is a continuous function $F^*(z)$ such that $F^*(z)/f_{z_0}(z)$ is continuous and nonzero in $N(z_0)$.

This is a famous result due to K. Oka [34] (comp. also K. Stein [47]). H. Cartan [18] and J. P. Serre [44] have reformulated this result in the language of *sheaves*, which was invented by J. Leray [30], and many further developments have been inspired by it (and the similar additive Cousin problem).

9.3 In the additive problem we prescribe poles. A *meromorphic* function is a function that can locally be written as the quotient of two holomorphic functions. In the additive problem the local functions are meromorphic and the compatibility condition is the following: $f_{z_j}(z) - f_{z_k}(z)$ *holomorphic in* $N(z_j) \cap N(z_k)$. The result is: The additive Cousin problem has no solution for general domains, but it can always be solved (without any further condition) for domains of holomorphy. This result again is due to K. Oka [33]. H. Cartan [18] has extended it to complex manifolds (Stein manifolds) and recently Grauert [24] has extended many of these results to "complex spaces."

10. ANALYTIC CONTINUATION, ENVELOPES OF HOLOMORPHY

10.1 A holomorphic function is determined entirely by its values in an arbitrarily small open set. Locally it can be developed into a power series. Such a power series development is called a "germ." The totality of germs forms a space, which for one variable is the *Riemann surface* of the function. In particular one may add "irregular" germs at branch points. The result is a manifold. For $n > 1$ one obtains a *complex manifold* if one does *not* include the irregular germs. If one *does* include them, then the germ-space, in general, is no longer a *manifold* (i.e. locally euclidean). Thus the full analogue of the Riemann surface is not the *complex manifold* but something more general called *complex space.* The theory of complex spaces is rather delicate. The interested reader will find a comprehensive treatment of complex spaces in Grauert [23] [24] and Grauert and Remmert [25].

10.2 In Sec. 5 we have described the phenomenon of simultaneous analytic continuation. Given a domain D that is not a domain of holomorphy. Then we may ask: *What is the largest domain $E(D)$ into which all functions that are holomorphic in D can be continued holomorphically?* It is not practical to limit ourselves here to schlicht domains (that is, open subsets of the C^n). Analytic continuation of functions of one variable may lead to *multiple-valued* functions, which may be made *single-valued* by considering the domain into which we continue, to be a (non-schlicht) Riemann surface. Similarly we may admit for the analytic continuation from a given domain $D \subset C^n$ non-schlicht domains. Thus if we admit complex manifolds as domains into which we continue, then the largest domain into which all functions that are holomorphic in D can be continued may be non-schlicht. Examples show that this possibility actually *does* occur. This largest domain of simultaneous holomorphic continuation is called "envelope of holomorphy of D." More information can be found in Bremermann [13].

10.3 The problem of constructing the envelope of holomorphy to a concretely given domain is rather difficult. It plays a role in applications. Information can be found in Bremermann [13]; comp. also Taylor [48], Langmaack [29], Stein [46], Bremermann, Oehme, Taylor [16] and Vladimirov [50].

Only for the tube domains (3.5) is the answer simple: *The envelope of holomorphy of a tube domain is its convex envelope* (Bochner-Martin [3], Bremermann [9]).

10.4 The following result follows directly from the definition; nevertheless it may be useful to state it:

If $E(D)$ is a domain such that every function that is holomorphic in D can be continued into $E(D)$ and if $E(D)$ is a domain of holomorphy, then $E(D)$ is the envelope of holomorphy of D.

REFERENCES

1. Behnke, H., and F. Sommer, *Theorie der analytischen Funktionen einer komplexen Veränderlichen.* Berlin-Göttingen-Heidelberg: Springer Verlag, 1955, 2nd ed. 1962.
2. Behnke, H., and P. Thullen, "Theorie der Funktionen mehrerer komplexer Veränderlichen," *Ergebn. d. Math.*, vol. 3, no. 3 (1934) (photostatically reprinted by Chelsea, New York).
3. Bochner, S., and W. T. Martin, *Several Complex Variables.* Princeton, N.J.: Princeton University Press, 1948.
4. Bergman, S., "Sur les fonctions orthogonales de plusieurs variables complexes avec les applications à la théorie des fonctions analytiques," *Mémor. Sci. Math.*, vol. 106 (Paris, 1947); and: "Sur la fonction noyau d'un domaine . . . ," *Mémor. Sci. Math.*, vol. 108 (Paris, 1948).
5. Bergman, S., "The kernel function and conformal mapping," *Mathem. Surv.*, no. 5 (New York, 1950). Second ed. in preparation.
6. Bergman, S., "Kernel function and extended classes in the theory of functions of complex variables," *Colloque s. l. fonctions de plusieurs var.*, Bruxelles, 1953, pp. 135–157.
7. Bers, Lipman, "Several complex variables," Lecture Notes, New York University, 1962–63.
8. Bremermann, H. J., "Die Charakterisierung von Regularitätsgebieten

durch pseudokonvexe Funktionen," *Schr. Math. Inst. Univ. Münster*, vol. 5 (1951).

9. Bremermann, H. J., "Die Holomorphiehüllen der Tuben- und Halbtubengebiete," *Math. Ann.*, vol. 117 (1954), pp. 406–423.

10. Bremermann, H. J., "Über die Äquivalenz der pseudokonvexen Gebiete und der Holomorphiegebiete im Raum von n komplexen Veränderlichen," *Math. Ann.*, vol. 128 (1954), pp. 63–91.

11. Bremermann, H. J., "Complex convexity," *Trans. Amer. Math. Soc.*, vol. 82 (1956), pp. 17–51.

12. Bremermann, H. J., "Die Charakterisierung Rungescher Gebiete durch plurisubharmonische Funktionen," *Math. Ann.*, vol. 136 (1958), pp. 173–186.

13. Bremermann, H. J., "Construction of the envelopes of holomorphy of arbitrary domains," *Rev. Mat. Hisp.-Amer.*, 4th ser., vol. 17 (1957), pp. 1–26.

14. Bremermann, H. J., "On a generalized Dirichlet problem for plurisubharmonic functions and pseudo-convex domains," *Trans. Amer. Math. Soc.*, vol. 91 (1959), pp. 246–276.

15. Bremermann, H. J., "Several complex variables," Lecture Notes, University of Washington, 1957.

16. Bremermann, H. J., R. Oehme, and J. G. Taylor, "Proof of dispersion relations in quantized field theories," *Phys. Rev.*, vol. 109 (1958), pp. 2178–2190.

17. Carathéodory, C., *Funktionentheorie*, vols. 1 and 2. Basel: Birkhäuser, 1950.

18. Cartan, H., "Variétés analytiques complexes et cohomologie," *Colloque s. l. fonctions de plusieurs var.*, Bruxelles, 1953, pp. 41–55.

19. Cartan, H., *Séminaire Éc. Norm. Sup.*, 1951–1952.

20. Cartan, H., and P. Thullen, "Regularitäts- und Konvergenzbereiche," *Math. Ann.*, vol. 106 (1932), pp. 617–647.

21. Fuks, B. A., *Introduction to the Theory of Analytic Functions of Several Complex Variables*, Moscow, Gosudarstv. Izdat. Fiz.-Mat. Lit., 1962. Translation from Russian by American Math. Soc., 1963.

22. Fuks, B. A., *Special Chapters from the Theory of Analytic Functions of Several Complex Variables*, Moscow, State Publ. House of Physics and Math., 1963 (in Russian). English translation: American Math. Soc., Providence, R.I., 1965.

23. Grauert, H., "Approximationssätze für holomorphe Funktionen mit Werten in komplexen Räumen," *Math. Ann.*, vol. 133 (1957), pp. 139–159.

24. Grauert, H., "Holomorphe Funktionen mit Werten in komplexen Lieschen Gruppen," *Math. Ann.*, vol. 133 (1957), pp. 450–472.

25. Grauert, H., and R. Remmert, "Komplexe Räume," *Math. Ann.*, vol. 136 (1958), pp. 245–318.

26. Hartogs, F., "Zur Theorie der anal. Funkt. mehr. unabhäng. Veränd. insbesondere über die Darstellung derselben durch Reihen welche nach Potenz. einer Veränderl. fortschreiten," *Math. Ann.*, vol. 62 (1906), pp. 1–88.

27. Hartogs, F., "Über die aus den singulären Stellen ein anal. Funkt. mehr. Veränd. bestehenden Gebilde," *Acta Math.*, vol. 32 (1909).

28. Hörmander, Lars, *An Introduction to Complex Analysis in Several Variables*. Van Nostrand, Princeton, N.J., 1966.

28a. Kallin, Eva, "Polynomial Convexity: Three Spheres Problem,' Proc. Conf. on Complex Analysis, Minneapolis, 1964. Springer-Verlag, 1965, pp. 301–304.

29. Langmaack, H., "Konstruktion von Holomorphiehüllen beliebiger Gebiete über dem C^n," Thesis, Münster, 1960.

30. Leray, J., "L'anneau spectral et l'anneau filtré d'homologie d'un espace localement compact et d'une application continue," and "L'homologie d'un espace fibré dont la fibre est connexe," *J. Math. Pures Appl.*, vol. 29 (1950), pp. 1–139 and pp. 169–213.

31. Norguet, F., "Sur les domaines d'holomorphie des fonctions uniformes de plusieurs variables complexes (Passage du local au global)," *Bull. Soc. Math. France,* vol. 82 (1954), pp. 137–159.

32. Oka, K., "Sur les fonctions de plusieurs variables I. Domaines convexes par rapport aux fonctions rationelles," *J. Sci. Hiroshima Univ.* Ser. A, vol. 6 (1936), pp. 245–255.

33. Oka, K., "— II. Domaines d'holomorphie," *J. Sci. Hiroshima Univ.* Ser. A, vol. 7 (1937), pp. 115–130.

34. Oka, K., "— III. Deuxième problème de Cousin," *J. Sci. Hiroshima Univ.* Ser. A, vol. 9 (1939), pp. 7–19.

35. Oka, K., "— IV. Domaines d'holomorphie et domaines rationellement convexes," *Japan J. Math.*, vol. 17 (1941), pp. 517–521.

36. Oka, K., "— V. L'intégrale de Cauchy," *Japan. J. Math.*, vol. 17 (1941), pp. 523–531.

37. Oka, K., "— VI. Domaines pseudoconvexes," *Tôhoku Math. J.*, vol. 49 (1942), pp. 15–52.

38. Oka, K., "— VII. Sur quelques notions arithmétiques," *Bull. Soc. Math. France*, vol. 78 (1950), pp. 1–27.

39. Oka, K., "— VIII. Lemme fondamental," *J. Math. Soc. Japan*, vol. 3 (1951), pp. 204–214 and pp. 259–278.

40. Oka, K., "— IX. Domaines finis sans point critique intérieur," *Japan J. Math.*, vol. 23 (1953), pp. 97–155.

41. Oka, K., *Sur les fonctions analytiques de plusieurs variables.* Tokyo: Iwanami Shoten, 1961.

42. Remmert, R., "Sur les espaces analytiques holomorphiquement séparables et holomorphiquement convexes," *Compt. rend. Ac. Sc.*, vol. 243 (1956), pp. 118–121.

43. Remmert, R., and K. Stein, "Über die wesentlichen singularitäten analytischer Mengen," *Math. Ann.*, vol. 126 (1953), pp. 263–306.

44. Serre, J. P., "Quelques problèmes globaux relatifs aux variétés de Stein," *Colloque s. l. fonctions de plusieurs var.*, Bruxelles, 1953, pp. 57–68.

45. Sommer, F., "Über die Integralformeln in der Funktionentheorie mehrerer komplexer Veränderlichen," *Math. Ann.*, vol. 125 (1952), pp. 172–182.

46. Stein, K., "Zur Theorie der Funktionen mehrerer komplexer Veränderlichen. Die Regularitätshüllen niederdimensionaler Mannigfaltigkeiten," *Math. Ann.*, vol. 114 (1937), pp. 543–569.

47. Stein, K., "Topologische Bedingungen für die Existenz analytischer Funktionen komplexer Veränderlichen zu vorgegebenen Nullstellenflächen," *Math. Ann.*, vol. 117 (1941), pp. 727–757.

48. Taylor, J., "A theorem of continuation for functions of several complex variables," *Proc. Cambr. Philos. Soc.*, vol. 54 (1958), pp. 377–382.

49. Titchmarsh, E. C., *The Theory of Functions*, 2nd ed. London-New York: Oxford University Press, 1939.

50. Vladimirov, V. S., *Metody Teorii Funktsii Mnogikh Kompleksnykh Peremennykh*, (Methods of the Theory of Functions of Many Complex Variables), *Navka*, Moscow, 1964. English translation: M.I.T. Press, Cambridge, Mass., 1966.

51. Weil, A., "L'intégrale de Cauchy et les fonctions de plusieurs variables," *Math. Ann.*, vol. 111 (1935).

NONLINEAR MAPPINGS BETWEEN BANACH SPACES

Lawrence M. Graves

1. INTRODUCTION

In the elementary theory of differential equations, we find a simple example of the kind of relationship between functions that we shall be considering. If $f(x, y)$ is a differentiable function of x and y in a domain D, and (x_0, y_0) is a point interior to D, then there is a unique function $y(x)$ defined on a suitably restricted interval $a \leq x \leq b$ and satisfying the conditions

$$\text{(1.1)} \qquad \frac{dy}{dx} = f[x, y(x)] \qquad \text{on } [a, b], \qquad y(x_0) = y_0.$$

When the point (x_0, y_0) and the function f are changed, the function y satisfying (1.1) may also change. Some properties of the relation between the function y and the point (x_0, y_0) are given in the standard existence theorems for differential equations. Properties of the relation between the function y and the function

f (extended to the case of systems of differential equations) were applied by G. A. Bliss [3] [4] to problems of calculating the effect of wind and other factors on the trajectories of projectiles.

The length L of the graph of a function $\phi(x)$ defined on an interval $[a, b]$ is a real-valued function $L(\phi)$ of the function ϕ. When ϕ has a continuous derivative, $L(\phi)$ is given by the familiar formula

$$L(\phi) = \int_a^b \sqrt{(1 + \phi'^2)}\, dx.$$

The area S of the surface generated by revolving the graph of ϕ about the x-axis is given by the formula

$$S(\phi) = 2\pi \int_a^b \phi\sqrt{(1 + \phi'^2)}\, dx.$$

These are special cases of integrals of the calculus of variations of the form

$$J(\phi) = \int_a^b f[x, \phi(x), \phi'(x)]\, dx, \tag{1.2}$$

where f is a fixed function of three variables, and a function ϕ is sought which will minimize (or maximize) the function J.

In case the integrand f depends also on a fourth variable t, then the formula

$$\psi(t) = \int_a^b f[t, x, \phi(x), \phi'(x)]\, dx + \phi(t) \tag{1.3}$$

defines a mapping T of a class of functions ϕ onto a class of functions ψ. Conditions can be given ensuring that T has an inverse mapping, defined at least locally. A theorem of this kind is a special type of implicit function theorem. We propose to consider a general implicit function theorem, which includes a wide variety of theorems as special cases. The methods used will involve approximation of nonlinear functions by linear functions. So we shall restrict attention to functions whose domain and ranges are contained in normed vector spaces, and which have properties of differentiability.

2. NORMED VECTOR SPACES, AND BANACH SPACES

A vector space $\mathfrak{Y}$ over a field $\mathfrak{A}$ consists of a class $\mathfrak{Y}$ of elements y (called vectors) and an operation $y_1 + y_2$ (called addition) with which $\mathfrak{Y}$ is a commutative group, together with an operation ay (called scalar multiplication), defined for all a in the field $\mathfrak{A}$ and all y, which is distributive with respect to addition of scalars (i.e., addition of numbers a and b in the field $\mathfrak{A}$) and with respect to addition of vectors, and which satisfies the associative law $a(by) = (ab)y$ and the condition $1 \cdot y = y$. The identity elements for the group $(\mathfrak{Y}, +)$ and for the additive group $(\mathfrak{A}, +)$ will be denoted by the same symbol 0, since it will always be clear from the context which is meant. For simplicity we shall restrict attention to the case when the field $\mathfrak{A}$ consists of all real numbers. However, there are many cases in applications where it is desirable to use the field of complex numbers.

A vector space $\mathfrak{Y}$ is said to be *normed* if there is a function $||y||$ defined on $\mathfrak{Y}$ with nonnegative real values, such that for all a, y, and y_1,

$$||y + y_1|| \leq ||y|| + ||y_1||,$$
$$||ay|| = |a| \, ||y||,$$
$$||y|| = 0 \qquad \text{only if } y = 0.$$

A normed vector space $\mathfrak{Y}$ is called *complete* in case every Cauchy sequence in $\mathfrak{Y}$ has a limit in $\mathfrak{Y}$, i.e., whenever $\lim_{m,n} ||y_m - y_n|| = 0$, there exists a point y in $\mathfrak{Y}$ such that $\lim_m ||y_m - y|| = 0$. A complete normed vector space is commonly called a *Banach space*, for the Polish mathematician Stefan Banach, whose book [1] did so much to establish the basic theory of such spaces.

Examples:

Some important examples of Banach spaces are the following.

1. The space $\mathfrak{E}_n$, which is the n-dimensional number space consisting of all n-tuples $(y^1, \cdots, y^n)$ of real numbers. Here we may take $||y|| = \max |y^i|$, or $||y|| = (\Sigma \, |y^i|^2)^{1/2}$, or $||y|| = \Sigma \, |y^i|$, without making any difference in the theory with which we shall be concerned.

2. Real Hilbert space $\mathfrak{H}$, consisting of all infinite sequences (y^i) of real numbers such that $||y|| = (\Sigma\, |y^i|^2)^{1/2}$ is finite.

3. The space $\mathfrak{L}_p$, corresponding to a fixed measurable set E and a number $p \geq 1$, and consisting of all functions $y(t)$ which are measurable on E and have $|y(t)|^p$ Lebesgue-integrable over E. Here we set $||y|| = \left[\int_E |y(t)|^p\, dt\right]^{1/p}$, and it becomes necessary to identify functions which are equal almost everywhere. Note that the subspace of $\mathfrak{L}_p$ consisting of all its members $y(t)$ which are continuous on the set E is a normed vector space, but generally fails to be complete.

4. The space $\mathfrak{C}$, composed of all functions $y(t)$ which are continuous and bounded on a fixed set E. Here we set $||y|| = \text{l.u.b. } |y(t)|$ on E. The set E is sometimes required to be bounded and closed, and may even be restricted to be an interval. The norm used in this example is generally called "the uniform norm," and it may be used in any class of bounded functions.

5. The space $\mathfrak{C}^{(n)}$, composed of all functions $y(t)$ which are continuous and bounded on a set E with suitable properties, and which have continuous and bounded derivatives up to and including those of the nth order. Here we may set $||y||$ equal to the least upper bound of $|y(t)|$ and of the absolute values of all its derivatives up to the nth order.

In all the examples listed, functions with complex values may be admitted. In that case, the field of scalars may consist of the complex numbers, or be restricted to the reals.

Examples of functions defined on the finite-dimensional spaces $\mathfrak{C}_n$ are familiar. The integrals $J(\phi)$ in (1.2) are defined on a subset of the space $\mathfrak{C}^{(1)}$, with real values. The formula (1.3) defines a mapping of a subset of $\mathfrak{C}^{(1)}$ into $\mathfrak{C}$, when the function f is continuous.

The formation of cartesian product spaces is familiar from plane analytic geometry, where the plane is treated as the cartesian product of two lines. The cartesian product $A \times B$ of two sets A and B is defined as the set of all ordered pairs (a, b) where a is in A and b is in B. For the case of two normed vector spaces $\mathfrak{Y}$ and

$\mathfrak{Z}$ with the same field of scalars, we may define a norm in the cartesian product $\mathfrak{Y} \times \mathfrak{Z}$ in a variety of ways. We shall use

$$||(y, z)|| = \max \{||y||, ||z||\}. \tag{2.1}$$

It is easy to see that

$$||(y, z)||_p = [||y||^p + ||z||^p]^{1/p}$$

is also a norm, provided $p \geq 1$, and that it gives the same class of open sets as (2.1).

As an example of a case for which there is frequent use, let $\mathfrak{Y}$ be the class of continuous functions y defined on an interval $[a, b]$, and let $\mathfrak{F}$ be the class of continuous functions f defined on the cartesian product of the interval $[a, b]$ with an interval $[c, d]$. Let $\mathfrak{Y}_0$ denote the subset of $\mathfrak{Y}$ for which $y(r)$ is in $[c, d]$ for every r in $[a, b]$. Then the equation

$$y_1(r) = f[r, y(r)]$$

defines a mapping $y_1 = G(y, f)$ defined on the cartesian product $\mathfrak{Y}_0 \times \mathfrak{F}$, with values in the space $\mathfrak{Y}$. If, instead, we take $\mathfrak{F}$ to be the class of continuous functions defined on $[a, b] \times [a, b] \times [c, d]$, then

$$y_2(r) = \int_a^r f[r, t, y(t)]\, dt$$

defines a mapping $y_2 = H(y, f)$ of $\mathfrak{Y}_0 \times \mathfrak{F}$ into $\mathfrak{Y}$, and

$$y_3(r) = \int_a^b f[r, t, y(t)]\, dt$$

defines another such mapping $y_3 = K(y, f)$. Each of these transformations is linear in f for each fixed function y in $\mathfrak{Y}_0$. (See Sec. 3 below.) If we use the uniform norm for the spaces $\mathfrak{Y}$ and $\mathfrak{F}$, it is also readily seen that each of these transformations is continuous in f uniformly with respect to y, since

$$||G(y, f)|| \leq ||f||,$$
$$||H(y, f)|| \leq (b - a)||f||,$$
$$||K(y, f)|| \leq (b - a)||f||.$$

Moreover, since each f is uniformly continuous on its domain,

G, H, and K are continuous in y for each f, and so are continuous in y and f together.

We also notice that if $\mathfrak{Y}_0$ is a subset of a Banach space $\mathfrak{Y}$, and $\mathfrak{W}$ is a Banach space, then the space of all continuous and bounded functions f mapping $\mathfrak{Y}_0$ into $\mathfrak{W}$ is a Banach space $\mathfrak{F}$, if we use the uniform norm, $||f|| = \text{l.u.b.}\ ||f(y)||$ for y in $\mathfrak{Y}_0$. The proof that $\mathfrak{F}$ is a normed linear space is exactly similar to that for the case of real-valued functions f, and so is the proof that if a sequence (f_n) satisfies the Cauchy condition $\lim_{m,n} ||f_m - f_n|| = 0$, then there exists a function f defined, bounded, and continuous on $\mathfrak{Y}_0$ such that $\lim_n ||f_n - f|| = 0$.

In the following sections, we shall use the German letters $\mathfrak{U}$, $\mathfrak{V}$, $\mathfrak{W}$, $\mathfrak{Y}$, $\mathfrak{Z}$ to denote Banach spaces, although in some of the theorems it will be evident that not all the spaces need to be complete.

3. SPACES OF LINEAR CONTINUOUS TRANSFORMATIONS

A mapping L of a space $\mathfrak{Y}$ into a space $\mathfrak{Z}$ is called *linear* in case

$$L(a_1y_1 + a_2y_2) = a_1L(y_1) + a_2L(y_2)$$

for all a_1, a_2, y_1, y_2. It is easy to verify that if a linear transformation L is continuous at one point (e.g. at $y = 0$) then it is continuous everywhere, and

$$||L|| = \text{l.u.b.}\ ||L(y)|| \qquad \text{for } ||y|| = 1$$

is finite. The collection of all such mappings L is readily seen to be a Banach space, which may be denoted by $\mathfrak{L}(\mathfrak{Y}, \mathfrak{Z})$.

In case a linear transformation L determines a one-to-one correspondence between $\mathfrak{Y}$ and a subset $\mathfrak{Z}_0$ of $\mathfrak{Z}$, then it is readily verified that its inverse transformation K, which maps $\mathfrak{Z}_0$ onto $\mathfrak{Y}$, is also linear. Such a K need not be continuous when L is, however. Consider the following example. Let $\mathfrak{Y}$ and $\mathfrak{Z}$ be the classical Hilbert space $\mathfrak{H}$, and for each $y = (y^i)$, let $L(y) = z = (y^i/i)$. Then L is linear, continuous, and one-to-one, but the sequence $y_n = (y_n^i)$ with $y_n^n = 1$, $y_n^i = 0$ for $i \neq n$, has $||y_n|| = 1$, while $||L(y_n)|| = 1/n$. In this case the subset $\mathfrak{Z}_0$, which is the range of L, is not closed, since it does not contain the point z with

$z^i = 1/i$. When $\mathfrak{Z}_0$ is closed, then it is itself a Banach space, and it may be shown that in this case the inverse K of L is necessarily continuous. (See p. 41, Theorem 5, of [1].)

When the spaces $\mathfrak{Y}$ and $\mathfrak{Z}$ are identical, we may define a composition (or product) of transformations, and powers of a transformation, in the usual way.

The following theorems are concerned with linear transformations having continuous inverses.

THEOREM 3.1: *Let L be a continuous linear transformation mapping the Banach space $\mathfrak{Y}$ into itself, and suppose $||L|| < 1$. Let I denote the identity transformation on $\mathfrak{Y}$. Then the transformation $T = I - L$ has the continuous inverse*

$$R = I + \sum_{n=1}^{\infty} L^n,$$

and

$$||R - I|| \leq \frac{||L||}{1 - ||L||}, \qquad ||R|| \leq \frac{1}{1 - ||L||}.$$

Proof: Set

$$R_k = I + \sum_{n=1}^{k} L^n.$$

Then $R_k(I - L) = (I - L)R_k = I - L^{n+1}$. Since $||L^{n+1}||$ tends to zero, the sequence R_k must have a limit R with $R(I - L) = (I - L)R = I$. Also

$$||R - I|| \leq \sum_{n=1}^{\infty} ||L||^n = \frac{||L||}{1 - ||L||}.$$

Let us now consider the space $\mathfrak{L}(\mathfrak{Y}, \mathfrak{Z})$ of all continuous linear transformations of $\mathfrak{Y}$ into $\mathfrak{Z}$, and let $\mathfrak{L}_0$ denote the subset consisting of all transformations T which map $\mathfrak{Y}$ *onto* $\mathfrak{Z}$ one-to-one. The subset $\mathfrak{L}_0$ may be empty, and is never a linear space, but it does have the following property.

THEOREM 3.2: *The subset $\mathfrak{L}_0$ is open in $\mathfrak{L}(\mathfrak{Y}, \mathfrak{Z})$. In particular, if T_0 has a continuous inverse R_0, and $||T - T_0|| \leq \alpha/||R_0||$, where $\alpha < 1$, then T has an inverse R with $||R - R_0|| \leq \alpha||R_0||/(1 - \alpha)$, $||R|| \leq ||R_0||/(1 - \alpha)$.*

Proof: If we set $L = R_0(T_0 - T)$, we see that L maps $\mathfrak{Y}$ into $\mathfrak{Y}$, and $||L|| \leq ||R_0|| \, ||T - T_0|| \leq \alpha$. By Theorem 3.1, $I - L$ has an inverse S mapping $\mathfrak{Y}$ onto $\mathfrak{Y}$, and $||S - I|| \leq \alpha/(1 - \alpha)$. Hence $R = SR_0$ maps $\mathfrak{Z}$ onto $\mathfrak{Y}$ one-to-one, and

$$I = S(I - L) = S(I - R_0(T_0 - T)) = SR_0T = RT.$$

Also $||R - R_0|| = ||(S - I)R_0|| \leq \alpha||R_0||/(1 - \alpha)$.

It is desirable in the applications to consider Banach spaces whose points are functions defined on an interval $[a, b]$ of the real axis, with values in a fixed Banach space $\mathfrak{B}$. Such a space $\mathfrak{Y}$ might consist of all bounded functions y on $[a, b]$ with values in $\mathfrak{B}$, where

$$(3.1) \qquad ||y|| = \text{l.u.b.} \, ||y(r)|| \qquad \text{on } [a, b].$$

If $y_1 = L(y_2)$, where L is a transformation of $\mathfrak{Y}$ into $\mathfrak{Y}$, we shall find it convenient to denote the value $y_1(r)$ by $L(y_2 \mid r)$.

THEOREM 3.3: *Let $\mathfrak{Y}$ be a Banach space of continuous functions on $[a, b]$ to $\mathfrak{B}$, with norm* (3.1), *and let L be a linear transformation mapping $\mathfrak{Y}$ into itself, and such that there exists a constant M such that*

$$(3.2) \qquad ||L(y \mid r)|| \leq M \int_a^r ||y(t)|| \, dt.$$

Then the transformation $T = I - L$ has a continuous inverse R with $||R|| \leq \exp{[M(b - a)]}$.

Proof: We wish to show that for each function z in $\mathfrak{Y}$, the equation

$$(3.3) \qquad y = z + L(y)$$

has a unique solution y. The proof is made by successive substitutions, as for Theorem 3.1, except that here we display the argument r of the various functions. For convenience set $y_0 = z$, $y_{n+1}(r) = z(r) + L(y_n \mid r)$ for $n = 0, 1, 2, \cdots$. Then we have

$$||y_1(r) - y_0(r)|| = ||L(y_0 \mid r)|| \leq M(r - a)||z||,$$

and by induction

$$(3.4) \qquad \|y_{n+1}(r) - y_n(r)\| = \|L(y_n - y_{n-1} \mid r)\| \leq M \int_a^r \|y_n(t) - y_{n-1}(t)\|\, dt \leq \frac{M^{n+1}(r-a)^{n+1}\|z\|}{(n+1)!}.$$

Thus since $\mathfrak{Y}$ is a Banach space, the sequence (y_n) converges to a function y in $\mathfrak{Y}$ (the convergence being uniform with respect to r), and since $L(y) - L(y_n) = L(y - y_n)$, $L(y_n)$ tends to $L(y)$, and so (3.3) has the solution y. By (3.4),

$$\|y\| \leq \|z\| \left[1 + \sum_{k=1}^{\infty} \frac{M^k(b-a)^k}{k!}\right] = \|z\| \exp\,[M(b-a)].$$

If there were two solutions, their difference η would satisfy the equation $\eta = L(\eta)$, and from (3.2) we would obtain by induction $\|\eta(r)\| \leq M^n(r-a)^n\|\eta\|/n!$, which implies $\eta = 0$.

4. DIFFERENTIALS OF FUNCTIONS

Let G be a function defined on an open set $\mathfrak{Y}_0$ in a normed vector space $\mathfrak{Y}$, with values in a normed vector space $\mathfrak{Z}$. Then G is said to be differentiable at a point y_1 in $\mathfrak{Y}_0$ in case there exists a linear continuous function L mapping $\mathfrak{Y}$ into $\mathfrak{Z}$, such that the quotient

$$\frac{\|G(y) - G(y_1) - L(y - y_1)\|}{\|y - y_1\|}$$

tends to zero with the denominator. It will be noted that this is a straightforward generalization of the usual definition for differentiability of a real-valued function of several real variables, and that the usual proof shows that a function G which is differentiable at y_1 is continuous at y_1. However, the continuity of the linear function L, which is required here, is in the finite-dimensional case a consequence of its linearity. The function L is called the differential of G at the point y_1. In case G has a differential at every y in $\mathfrak{Y}_0$, we shall use the notation $dG(y; \delta y)$ for its value at the point y and the vector δy. (The semicolon is used to indicate the special role of the variable δy.) Thus dG is a function defined on the cartesian product $\mathfrak{Y}_0 \times \mathfrak{Y}$, with values in the space $\mathfrak{Z}$.

Alternately, dG may be regarded as a function defined on $\mathfrak{Y}_0$ with values in the space $\mathfrak{L}(\mathfrak{Y}, \mathfrak{Z})$ of linear continuous transformations of $\mathfrak{Y}$ into $\mathfrak{Z}$. From this point of view, $dG(y)$ is sometimes called the derivative of G at y.

It is easily verified that when G is differentiable at a point y, then for each δy in $\mathfrak{Y}$

$$(4.1) \qquad dG(y; \delta y) = \frac{d}{d\alpha} G(y + \alpha\delta y) \qquad \text{at } \alpha = 0.$$

However, the right-hand side of (4.1) may exist and be a continuous linear function of δy for functions G which are not differentiable. For example, let $G(v, w) = vw^2(v^2 + w^4)^{-3/4}$, where v and w are real variables, and $G(0, 0) = 0$. Then G is continuous, and

$$\frac{d}{d\alpha} G(\alpha v, \alpha w) = 0 \qquad \text{at } \alpha = 0.$$

However, if we take $v = w^2$, we find

$$\frac{|G(v, w)|}{(v^2 + w^2)^{1/2}} = 2^{-3/4}(w^2 + 1)^{-1/2},$$

so G is not differentiable at $(0, 0)$.

When dG is continuous as a function on $\mathfrak{Y}_0$ with values in $\mathfrak{L}(\mathfrak{Y}, \mathfrak{Z})$, the original function G is said to be of class C' on $\mathfrak{Y}_0$. Put into ϵ, δ terms, the definition of the continuity required of dG reads as follows: for every y_1 in $\mathfrak{Y}_0$ and every $\epsilon > 0$ there exists $\delta_\epsilon > 0$ such that, when y is in $\mathfrak{Y}_0$, $||y - y_1|| < \delta_\epsilon$, and $||\delta y|| = 1$, then $||dG(y; \delta y) - dG(y_1; \delta y)|| < \epsilon$. It is easy to see that for a fixed domain $\mathfrak{Y}_0$, the class C' of functions is a vector space, and the operator d is a linear operator on C'. For example, if $G(y) = G_1(y) + G_2(y)$, then

$$\begin{aligned} G(y) - G(y_1) - dG_1(y_1; y - y_1) - dG_2(y_1; y - y_1) \\ = G_1(y) - G_1(y_1) - dG_1(y_1; y - y_1) + G_2(y) \\ - G_2(y_1) - dG_2(y_1; y - y_1). \end{aligned}$$

Upon dividing by $||y - y_1||$ and taking the limit as $||y - y_1||$ tends to zero, we find $dG = dG_1 + dG_2$. When the space $\mathfrak{Y}$ is the cartesian product of two spaces $\mathfrak{V}$ and $\mathfrak{W}$, the differential is the sum of partial differentials. For, since differentials are linear functions,

$$(4.2) \qquad dG(y; \delta y) = dG(y; (\delta v, 0)) + dG(y; (0, \delta w)),$$

where y is the pair (v, w) and δy is the pair $(\delta v, \delta w)$. When w is fixed, i.e. G is regarded as a function of v alone, then $dG(y; (\delta v, 0))$ is obviously the differential of G, and it is natural to use for it the notation $d_v G(y; \delta v)$, and to write

$$(4.3) \qquad dG(y; \delta y) = d_v G(y; \delta v) + d_w G(y; \delta w)$$

in place of (4.2). The fact is familiar from the case of functions of two real variables that a function may have partial differentials without being differentiable as a function of the two variables together. Later (in Sec. 5) we shall be able to prove that when the partial differentials $d_v G(y; \delta v)$ and $d_w G(y; \delta w)$ are continuous functions of y uniformly for $||\delta v|| = ||\delta w|| = 1$, then G is of class C' as a function of y.

5. INTEGRALS OF VECTOR-VALUED FUNCTIONS OF A REAL VARIABLE

If $F(r)$ is defined for r on the real interval $[a, b]$, with values in a bounded subset of a Banach space $\mathfrak{Y}$, then Riemann sums may be defined for it in the usual way. The limit of these sums, as the norm of the partition tends to zero, if it exists, may be called the Riemann integral of F and denoted by

$$\int_a^b F(r)\, dr.$$

It is readily seen that when F is continuous, the norm of the difference of two sums will be less than an arbitrary $\epsilon > 0$ when the norms of their partitions are sufficiently small, so in this case the limit of the sums will surely exist, since $\mathfrak{Y}$ is assumed to be complete. However, there are cases, when the space $\mathfrak{Y}$ has infinite dimensionality, in which an everywhere discontinuous function exists which is integrable in the sense just described. (See [7] for an example.) Many other generalizations of the notion of integral to vector-valued functions have been developed, but they will not be discussed here.

When a function $G(r)$ of a real variable is differentiable at r_0, $dG(r_0; \delta r)$ equals the product of δr by a vector $G'(r_0)$ which is appropriately called the derivative of G at r_0. Now suppose $G(r)$ has a derivative $G'(r)$ at each point of $[a, b]$, and that $G'(r)$ is integrable on $[a, b]$. Then the fundamental theorem of integral calculus holds, that is,

$$G(b) - G(a) = \int_a^b G'(r)\, dr. \tag{5.1}$$

This may be shown as follows. Take an arbitrary $\epsilon > 0$. Then there exists $\delta > 0$ such that for every partition P of $[a, b]$ by points r_i, with norm $N(P) < \delta$,

$$||\sum_P G'(\sigma_i)\Delta_i - \int_a^b G'(r)\, dr|| < \epsilon,$$

where $\Delta_i = r_{i+1} - r_i$, and $r_i \leq \sigma_i \leq r_{i+1}$. From the definition of derivative, for each point r of $[a, b]$, there exists a positive number $\alpha_r < \delta$ such that if $|r' - r| \leq \alpha_r$, then

$$||G(r') - G(r) - G'(r)(r' - r)|| \leq \epsilon|r' - r|. \tag{5.2}$$

By the Borel theorem, a finite number of the open intervals $(r - \alpha_r, r + \alpha_r)$ cover $[a, b]$. Let these be denoted by $I_1, \cdots, I_m$, and their centers by $\rho_1, \cdots, \rho_m$. We may suppose that no interval I_k is included in the union of the remaining I_j's, and that $\rho_1 < \rho_2 < \cdots < \rho_m$. Then for the points r_i of the partition P we take the points a and b and all the points ρ_k, and also the right-hand end-point of each interval I_k which does not contain ρ_{k+1}. So $N(P) < \delta$, and there is at least one end-point of each interval $[r_i, r_{i+1}]$ which is one of the ρ_k, and which we denote by σ_i. Thus we have from (5.2)

$$||G(r_{i+1}) - G(r_i) - G'(\sigma_i)\Delta_i|| \leq \epsilon\Delta_i.$$

Hence

$$\begin{aligned} ||G(b) - G(a) - \int_a^b G'(r)\, dr|| & \\ &\leq \Sigma\, ||G(r_{i+1}) - G(r_i) - G'(\sigma_i)\Delta_i|| \\ &\quad + ||\Sigma\, G'(\sigma_i)\Delta_i - \int_a^b G'(r)\, dr|| \\ &< \epsilon(b - a + 1). \end{aligned}$$

From (5.1) we may readily derive the following substitute for the theorem of mean value, which will be useful later.

THEOREM 5.1: *Suppose the function H is of class C' on a convex open subset $\mathfrak{Y}_0$ of the space $\mathfrak{Y}$, with values in the complete space $\mathfrak{Z}$. Then for each pair of points y_1, y_2 in $\mathfrak{Y}_0$,*

$$H(y_2) - H(y_1) = \int_0^1 dH[y_1 + r(y_2 - y_1); y_2 - y_1]\, dr.$$

To verify this, it suffices to set $G(r) = H[y_1 + r(y_2 - y_1)]$, since then $G'(r) = dH[y_1 + r(y_2 - y_1); y_2 - y_1]$, which is continuous and hence integrable with respect to r.

From the preceding theorem we can derive the following result, to which reference was made in Sec. 4.

THEOREM 5.2: *Let $\mathfrak{Y}$ denote the cartesian product $\mathfrak{V} \times \mathfrak{W}$, and suppose the function G maps the open subset $\mathfrak{Y}_0$ of $\mathfrak{Y}$ into the space $\mathfrak{Z}$ and has partial differentials $d_vG(y; \delta v)$ and $d_wG(y; \delta w)$ which are continuous in y uniformly for $||\delta v|| = ||\delta w|| = 1$. Then G is of class C' on $\mathfrak{Y}_0$, and $dG(y; \delta v, \delta w) = d_vG(y; \delta v) + d_wG(y; \delta w)$.*

Proof: By Theorem 5.1 and the continuity of d_vG we have

$$\begin{aligned} &||G(v, w) - G(v_1, w) - d_vG(v_1, w_1; v - v_1)|| \\ &\quad = ||\int_0^1 [d_vG(v_1 + t(v - v_1), w; v - v_1) - d_vG(v_1, w_1; v - v_1)]\, dt|| \\ &\quad \leq \epsilon||v - v_1|| \end{aligned}$$

when $||v - v_1||$ and $||w - w_1||$ are sufficiently small. Also

$$||G(v_1, w) - G(v_1, w_1) - d_wG(v_1, w_1; w - w_1)|| \leq \epsilon||w - w_1||$$

when $||w - w_1||$ is sufficiently small. Hence

$$\begin{aligned} ||G(v, w) - G(v_1, w_1) - d_vG(v_1, w_1; v - v_1) - d_wG(v_1, w_1; w - w_1)|| \\ \leq 2\epsilon||y - y_1|| \end{aligned}$$

when we take $||y - y_1|| = \max \{||v - v_1||, ||w - w_1||\}$ sufficiently small. Thus $dG(y; \delta v, \delta w) = d_vG(y; \delta v) + d_wG(y; \delta w)$, and the required continuity of dG is immediate.

6. BASIC IMPLICIT FUNCTION THEOREMS

The first theorem is an extension of Theorem 3.1 to the case of a nonlinear function. Here $\mathfrak{Y}$ and $\mathfrak{Z}$ denote arbitrary Banach spaces.

THEOREM 6.1: *Let $F(y)$ be defined for $||y - y_0|| < a$, with values in the space $\mathfrak{Y}$, and suppose there is a constant $k < 1$ such that*

$$||F(y_1) - F(y_2)|| \leq k||y_1 - y_2||$$

whenever $||y_1 - y_0|| < a$, $||y_2 - y_0|| < a$, and

$$||F(y_0) - y_0|| < (1 - k)a.$$

Then there exists a unique point y such that $y = F(y)$ and $||y - y_0|| < a$.

Proof: The proof is by the method of successive substitutions. Set

$$y_n = F(y_{n-1}) \qquad \text{for } n = 1, 2, \cdots.$$

Then $||y_{n+1} - y_n|| \leq k\, ||y_n - y_{n-1}||$, and by induction

$$||y_{n+1} - y_n|| \leq k^n||y_1 - y_0||,$$

$$||y_n - y_0|| \leq (1 + k + \cdots + k^{n-1})||y_1 - y_0||$$

$$< \frac{||y_1 - y_0||}{1 - k} < a.$$

Thus all the approximations y_n are well defined, and since the space $\mathfrak{Y}$ is complete, the sequence (y_n) converges to a point y with $||y - y_0|| \leq ||y_1 - y_0||/(1 - k) < a$. Since F is continuous, $F(y_{n-1})$ converges to $F(y)$. The uniqueness of the solution with $||y - y_0|| < a$ is readily verified.

The next theorem gives sufficient conditions for the existence of a solution of an equation $G(y) = 0$. The two theorems with their proofs constitute an extension of Newton's method.

THEOREM 6.2: *Let $G(y)$ be defined for $||y - y_0|| < a$ with values in the space $\mathfrak{Y}$, and let T be a continuous linear transformation mapping $\mathfrak{Y}$ onto itself and having a continuous inverse R. Let M be a positive constant such that whenever $||y - y_0|| < a$, $||y_1 - y_0|| < a$, then*

$$(6.1) \qquad ||G(y) - G(y_1) - T(y - y_1)|| \leq M||y - y_1||,$$

$$(6.2) \qquad M||R|| < 1,$$

$$(6.3) \qquad ||R|| \, ||G(y_0)|| < (1 - M||R||).$$

Then there exists a unique point y such that $||y - y_0|| < a$ and $G(y) = 0$.

Proof: If we set

$$F(y) = y - R[G(y)]$$

then the equation $y = F(y)$ is equivalent to $G(y) = 0$. Thus the conclusion follows from Theorem 6.1, since by (6.1),

$$||F(y) - F(y_1)|| = ||R[T(y - y_1) - G(y) + G(y_1)]|| \leq M||R|| \, ||y - y_1||,$$

and by (6.3)

$$||F(y_0) - y_0|| = ||R[G(y_0)]|| < (1 - M||R||).$$

The next theorem is a straightforward generalization of the elementary implicit function theorem.

THEOREM 6.3: *Let $G(y, z)$ be defined and of class C' on an open set S in the cartesian product of $\mathfrak{Y}$ and $\mathfrak{Z}$, with values in a Banach space $\mathfrak{W}$, and let (y_0, z_0) be a point of S such that $G(y_0, z_0) = 0$. Suppose also that the linear transformation T_0 defined by*

$$(6.4) \qquad T_0(\delta y) = d_y G(y_0, z_0; \delta y)$$

maps $\mathfrak{Y}$ onto $\mathfrak{W}$ one-to-one. Then there exist positive constants a and b and a function $Y(z)$ defined and of class C' for $||z - z_0|| < b$ such that for $||y - y_0|| < a$, $||z - z_0|| < b$, we have: (i) *(y, z) is in S;* (ii) *$||Y(z) - y_0|| < a$;* (iii) *$G[Y(z), z] = 0$;* (iv) *if $G(y, z) = 0$ then $y = Y(z)$;* (v) *the linear transformation $T(\delta y) = d_y G[Y(z), z; \delta y]$ maps $\mathfrak{Y}$ onto $\mathfrak{W}$ one-to-one.*

Proof: We shall make the proof with the initial condition $G(y_0, z_0) = 0$ replaced by the weaker hypothesis $||G(y_0, z_0)|| < 1/||R_0||$, where R_0 is the inverse of the linear transformation T_0 given in (6.4). Take $\epsilon > ||G(y_0, z_0)||$, $\epsilon < 1/||R_0||$. By Theorem 3.2, if $||T - T_0|| \leq \epsilon$, then T has an inverse R with

$$(6.5) \qquad ||R|| \leq \frac{||R_0||}{1 - \epsilon||R_0||}.$$

Since G is of class C', for $0 < \beta < (1/\|R_0\| - \epsilon)$ there is a positive constant a such that when $\|y - y_0\| < a$, $\|y_1 - y_0\| < a$, $\|z - z_0\| < a$, we have

$$\|d_yG(y, z; \delta y) - d_yG(y_0, z_0; \delta y)\| \leq \beta\|\delta y\|,$$

$$\left\|\int_0^1 d_yG(y_1 + t(y - y_1), z; \delta y)\, dt - d_yG(y_0, z_0; \delta y)\right\| \leq \beta\|\delta y\|,$$

and hence by Theorem 5.1

$$\|G(y, z) - G(y_1, z) - d_yG(y_0, z_0; y - y_1)\| \leq \beta\|y - y_1\|.$$

Also by Theorem 3.2, the transformations $d_yG(y, z; \delta y)$ and

$$\int_0^1 d_yG(y_1 + t(y - y_1), z; \delta y)\, dt$$

have inverses with norm $\leq \|R_0\|/(1 - \beta\|R_0\|) < 1/\epsilon$. Thus for each z the hypotheses (6.1) and (6.2) of Theorem 6.2 are fulfilled with T replaced by T_0, R by R_0, and $M = 1/\|R_0\| - \epsilon$. Since G is continuous in z and $\|G(y_0, z_0)\| < \epsilon$, there exists a positive $b \leq a$ such that when $\|z - z_0\| < b$, then $\|G(y_0, z)\| < \epsilon$, and so for these values of z the hypothesis (6.3) is also satisfied. Hence Theorem 6.2 yields the statements (i), (ii), $\cdots$, (v). It remains to prove that the function $Y(z)$ is of class C' for $\|z - z_0\| < b$.

We first show that $Y(z)$ is continuous at each point z_1 with $\|z_1 - z_0\| < b$, as follows. Letting Y_1 denote $Y(z_1)$, we have

$$\begin{aligned} 0 &= G(Y, z) - G(Y_1, z) + G(Y_1, z) - G(Y_1, z_1) \\ &= \int_0^1 d_yG(Y_1 + t(Y - Y_1), z; Y - Y_1)\, dt + G(Y_1, z) - G(Y_1, z_1). \end{aligned}$$

If we denote the inverse of

$$\int_0^1 d_yG(Y_1 + t(Y - Y_1), z; \delta y)\, dt$$

by R_z, we find

$$Y - Y_1 = -R_z[G(Y_1, z) - G(Y_1, z_1)], \tag{6.6}$$

and since G is continuous in z when y is fixed, and $\|R_z\| < 1/\epsilon$, the continuity of the function Y follows. Now let R_1 denote the inverse of $d_yG(Y_1, z_1; \delta y)$. We shall show that $-R_1d_zG(Y_1, z_1; \delta z)$ is the differential of the function Y at the point z_1. From (6.6) we have

$$\begin{aligned}\Delta &= Y - Y_1 + R_1 d_z G(Y_1, z_1; z - z_1) \\ &= -R_z[G(Y_1, z) - G(Y_1, z_1) - d_z G(Y_1, z_1; z - z_1)] \\ &\quad - (R_z - R_1) d_z G(Y_1, z_1; z - z_1).\end{aligned}$$

Now by the differentiability of G, when $||z - z_1||$ is sufficiently small,

$$||G(Y_1, z) - G(Y_1, z_1) - d_z G(Y_1, z_1; z - z_1)|| \leq \delta ||z - z_1||,$$

and by the continuity of $d_y G$ and Y and Theorem 3.2, $||R_z - R_1|| < \delta$. Also by the first part of the proof, $||R_z|| < 1/\epsilon$. Hence

$$||\Delta|| \leq \frac{\delta ||z - z_1||}{\epsilon} + \delta ||d_z G|| \, ||z - z_1||,$$

so that Y is differentiable at z_1. The continuity of $dY(z_1; \delta z) = -R_1 d_z G(Y_1, z_1; \delta z)$ in z_1 is readily verified with the help of Theorem 3.2.

7. APPLICATIONS TO DIFFERENTIAL EQUATIONS

An initial-value problem for an ordinary differential equation of the first order may be written in the form

$$f[r, v'(r), v(r)] = 0, \qquad v(a) = x. \tag{7.1}$$

If a solution $v_0(r)$ of the system (7.1), which is of class C' on a closed interval $[a, b]$, is known, with $v_0(a) = x_0$, it is frequently desirable to know that a solution $v(r, x)$ is defined for $a \leq r \leq b$ and for each x near x_0, and that the function $v(r, x)$ is of class C'. This result may be obtained by first applying an extended implicit function theorem to solve the equation $f(r, v', v) = 0$ for v' near $v_0'(r)$, and then applying the ordinary embedding theorem for differential equations. (See, e.g., pp. 144 and 163 ff. of [10].) It is our purpose here to show that Theorem 6.3 may be applied directly to obtain the desired result, with the help of Theorem 3.3. In place of the system (7.1) we shall consider a more general system, in order to give a slightly broader idea of the scope of Theorem 6.3.

In place of $v'(r)$ we shall write $y(r)$. Then

$$v(r) = x + \int_a^r y(t)\, dt,$$

and (7.1) becomes

$$f[r, y(r), x + \int_a^r y(t)\,dt] = 0. \tag{7.2}$$

We now let $\mathfrak{Y}$ denote the Banach space of all real-valued continuous functions defined on the interval $[a, b]$, with the uniform norm, and let $h(r, y, x)$ be a real-valued function continuous for r in $[a, b]$, y in an open set $\mathfrak{Y}_0$ in the space $\mathfrak{Y}$, and x in an open interval (c, d). We suppose also that the function h is of class C' in (y, x) with differentials $d_y h(r, y, x; \delta y)$ and $d_x h(r, y, x; \delta x)$ continuous in (r, y, x). Let y_0 be an element of $\mathfrak{Y}_0$, and let $f(r, u, v, x)$ be a real-valued function continuous for r in $[a, b]$, x in (c, d), $|u - y_0(r)| < \epsilon$, $|v - h(r, y_0, x)| < \epsilon$, and of class C' in (u, v, x) with partial derivatives continuous in (r, u, v, x). Then in place of (7.2) above we consider the equation

$$f[r, y(r), h(r, y, x), x] = 0 \qquad \text{for } a \leq x \leq b. \tag{7.3}$$

THEOREM 7.1: *In addition to the conditions stated above on the functions h, f, and y_0, suppose that*

$$f[r, y_0(r), h(r, y_0, x_0), x_0] = 0 \qquad \textit{on } [a, b], \tag{7.4}$$

where x_0 is in (c, d), and that the partial derivative

$$f_u[r, y_0(r), h(r, y_0, x_0), x_0] \neq 0 \qquad \textit{on } [a, b]. \tag{7.5}$$

Suppose also that there is a constant k such that

$$|d_y h(r, y_0, x_0; \delta y)| \leq k \int_a^r |\delta y(t)|\,dt \qquad \textit{on } [a, b]. \tag{7.6}$$

Then there exist positive constants α and β and a function $Y(x)$ defined for $|x - x_0| < \beta$, such that $\alpha \leq \epsilon$ and:

(i) $||Y(x) - y_0|| < \alpha$;

(ii) $f[r, Y(x \mid r), h(r, Y(x), x), x] = 0$ *for* $a \leq r \leq b$, $|x - x_0| < \beta$;

(iii) *if* $|x - x_0| < \beta$, $||y - y_0|| < \alpha$, *and* (7.3) *holds, then* $y = Y(x)$;

(iv) $Y(x \mid r)$ *has a partial derivative with respect to x which is continuous in (x, r).*

Proof: Let k_1 be the greatest lower bound of $|f_u[r, y_0(r), h(r, y_0, x_0), x_0]|$ on $a \leq r \leq b$. Then $k_1 > 0$ by (7.5). If we set

$$G(x, y \mid r) = f[r, y(r), h(r, y, x), x], \tag{7.7}$$

it is a straightforward application of the definitions to show that $G(x, y)$ is of class C' for x near x_0 and y near y_0, with values in the space $\mathfrak{Y}$. In order to apply Theorem 6.3, it suffices to show that the linear transformation $d_yG(x_0, y_0; \delta y)$ has an inverse defined on the whole of $\mathfrak{Y}$. We have

$$d_yG(x_0, y_0; \delta y \mid r) = f_u\, \delta y(r) + f_v\, d_yh(r, y_0, x_0; \delta y),$$

where the arguments of the partial derivatives f_u and f_v are $[r, y_0(r), h(r, y_0, x_0), x_0]$. The linear transformation $\delta z(r) = K_0(\delta y \mid r) = f_u\, \delta y(r)$ has an inverse $L_0(\delta z \mid r) = \delta z(r)/f_u$, so the equation $d_yG(x_0, y_0; \delta y) = \delta z$ is equivalent to

$$\delta y(r) + \frac{f_v d_y h(r, y_0, x_0; \delta y)}{f_u} = \frac{\delta z(r)}{f_u}. \tag{7.8}$$

There is a constant k_2 such that

$$|f_v[r, y_0(r), h(r, y_0, x_0), x_0]| \leq k_2,$$

so by (7.6),

$$\left| \frac{f_v\, d_y h(r, y_0, x_0; \delta y)}{f_u} \right| \leq \frac{k_2 k}{k_1} \int_a^r |\delta y(t)|\, dt.$$

Thus by Theorem 3.3, the equation (7.8) has a unique solution $\delta y = LL_0(\delta z)$, where L and hence LL_0 are linear and continuous. So the desired conclusion follows from Theorem 6.3.

Theorem 7.1 has an immediate extension to systems of equations, and in fact to the case when the variable u and the values of the functions y and f are in a Banach space $\mathfrak{U}$, v and the values of h are in a Banach space $\mathfrak{V}$, and x is in a Banach space $\mathfrak{X}$.

In conclusion we observe that from Theorem 6.1 we may obtain a short proof of the existence of a solution of an ordinary differential equation (cf. Birkhoff and Kellogg [2]). Let $f(r, v)$ be continuous for $a \leq r \leq b$, $|v - v_0| \leq c$, and satisfy a Lipschitz condition with respect to v, with constant k. Let $M = \max |f(r, v_0)|$, and let ϵ be a positive number less than each of the quantities $(b - a)$, $c/2M$, and $1/2k$. Let $\mathfrak{Y}$ denote the space of all real continuous functions $y(r)$ on the interval $[a, a + \epsilon]$, and let $\mathfrak{Y}_0$ be the subset for which $|y(r) - v_0| < c$. Then set

$$F(y \mid r) = v_0 + \int_a^r f(t, y(t))\, dt,$$

$$y_0(r) = v_0 \qquad \text{on } a \leq r \leq a + \epsilon.$$

The function F is defined on $\mathfrak{Y}_0$ and has values in $\mathfrak{Y}$, and

$$||F(y_1) - F(y_2)|| \leq k\epsilon \, ||y_1 - y_2|| < \frac{||y_1 - y_2||}{2},$$

$$||F(y_0) - y_0|| \leq M\epsilon < \frac{c}{2},$$

so Theorem 6.1 gives the existence of a unique solution of the equation $y = F(y)$, which is equivalent to $y'(r) = f(r, y)$ on $[a, a + \epsilon]$, $y(a) = v_0$.

REFERENCES

1. Banach, S., *Théorie des opérations linéaires.* Warsaw, and Chelsea, New York, 1932.
2. Birkhoff, G. D., and O. D. Kellogg, "Invariant points in function space," *Trans. Amer. Math. Soc.*, vol. 23 (1922), pp. 96–118.
3. Bliss, G. A., "Functions of lines in ballistics," *Trans. Amer. Math. Soc.*, vol. 21 (1920), pp. 93–106.
4. Bliss, G. A., *Mathematics for Exterior Ballistics.* New York, 1944.
5. Dunford, N., and J. T. Schwarz, *Linear Operators.* Interscience, New York, 1958.
6. Evans, G. C. *Functionals and Their Applications.* American Mathematical Society, 1916.
7. Graves, L. M., "Riemann integration and Taylor's theorem in general analysis," *Trans. Amer. Math. Soc.*, vol. 29 (1927), pp. 163–177.
8. Graves, L. M., "Implicit functions and differential equations in general analysis," *Trans. Amer. Math. Soc.*, vol. 29 (1927), pp. 514–552.
9. Graves, L. M., "Topics in the functional calculus," *Bull. Amer. Math. Soc.*, vol. 41 (1935), pp. 641–662.
10. Graves, L. M., *Theory of Functions of Real Variables.* McGraw-Hill, New York, 1946, 1956.

11. Hildebrandt, T. H., and L. M. Graves, "Implicit functions and their differentials in general analysis," *Trans. Amer. Math. Soc.*, vol. 29 (1927), pp. 127–153.

12. Riesz, F., and B. Sz.-Nagy, *Functional Analysis*. Stechert-Hafner, New York, 1955.

13. Volterra, V., *Leçons sur les fonctions de lignes*. Gauthier-Villars, Paris, 1913.

WHAT IS A SEMI-GROUP?

Einar Hille

1. INTRODUCTION

Like Monsieur Jourdain in *Le Bourgeois Gentilhomme*, who found to his great surprise that he had spoken prose all his life, mathematicians are becoming aware of the fact that they have used semi-groups extensively even if not always consciously. Their belated realization had good reasons. The concept was formulated and named as recently as 1904, and it is such a primitive notion that one may well be in doubt concerning its value and possible implications.

A semi-group S is an associative groupoid.

In other words, S is a collection of elements a, b, c, $\cdots$ together with a *binary operation* to be denoted by $\circ$ such that $a \circ b$ is uniquely defined as an element of S for every ordered pair of elements a, b of S and

$$a \circ (b \circ c) = (a \circ b) \circ c.$$

We can think of the operation as a mapping of the product set $S \times S$ into S by a function $F(a, b)$ which has a value in S if a and b are elements of S. Now the associativity is expressed by the functional equation

$$F(a, F(b, c)) = F(F(a, b), c).$$

The reader will recognize that these are properties of a group. It should be noted, however, that a group has other properties which are not postulated for a semi-group. Thus we do not assume the existence of a *unit element* e such that

$$a \circ e = e \circ a = a$$

for all a in S. Many semi-groups have such a unit element, but this is not required in general. Similarly, some elements of a semi-group with unit element may have *inverses*, but this is not postulated as a general property.

A semi-group is said to be *commutative* or *abelian* if

$$a \circ b = b \circ a.$$

If the operation is that of addition, we also speak of a *semi-module*.

When semi-groups were first introduced by J. A. de Séguier he also postulated a *law of cancellation:*

If $a \circ b = a \circ c$ *(or* $b \circ a = c \circ a$*) for all* a *in* S, *then* $b = c$.

We shall not make this assumption in the following. Any semi-group with a finite number of elements satisfying this law is actually a group.

So far nothing has been said about the set S. What we have defined is an *abstract semi-group*. By giving S additional structure, we open the possibility of obtaining interesting mathematical theories. There are several possibilities among which we mention the following.

S may be a subset of

(1) a *ring*, or of
(2) a *topological space*, or of
(3) an *algebra of operators*.

We can then speak of *algebraic* or *topological* or *transformation semi-groups* according to the character of S.

A ring is a semi-group under either of the basic operations of the ring which are usually referred to as addition and multiplication. This brings out the fact that the same set of elements may very well define a semi-group under several distinct operations.

To illustrate this point further, let us consider the set E^+ of positive real numbers. Suppose that $F(a, b)$ is a function of two variables with the following properties:

(i) $F(a, b)$ is defined as a positive number if a and b are positive.
(ii) $F(a, F(b, c)) = F(F(a, b), c)$ for all $a, b, c > 0$.

Then E^+ is a semi-group under the operation defined by

$$a \circ b = F(a, b).$$

There is no lack of functions $F(a, b)$ with these properties. A few examples are given by

$$a + b, \quad ab, \quad \frac{a+b}{1+ab}, \quad a(1 + b^2)^{1/2} + b(1 + a^2)^{1/2}.$$

The reader will recognize that these functions are related to the addition theorems of the functions

$$x, \qquad e^x, \qquad \tanh x, \qquad \sinh x,$$

respectively.

A topological semi-group is a topological space which is also a semi-group under an operation $\circ$ *such that* $a \circ b$ *is a continuous function of* a *and* b *in the topology of the space.* To make this precise, suppose that the topology of S is determined by a system of neighborhoods which satisfy the axioms of Hausdorff. We then require that to every neighborhood $N(a \circ b)$ of $a \circ b$ there is a pair of neighborhoods $N(a)$ of a and $N(b)$ of b such that

$$a \circ N(b) \subset N(a \circ b), \qquad N(a) \circ b \subset N(a \circ b).$$

Transformation semi-groups usually arise in the following manner. Let X be a Banach space, that is, a complete normed linear vector space. A linear bounded transformation of X into itself is a mapping $x \longrightarrow Tx$ which assigns to each element x of X a definite element $y = Tx$ of X subject to the two conditions

(i) $T[\alpha x_1 + \beta x_2] = \alpha T x_1 + \beta T x_2$,

(ii) $||Tx|| \leq M||x||$.

The first condition expresses that T is linear, the second that T is bounded. The set of all such transformations forms an *operator algebra* $E[X]$ for it contains sums, scalar multiples, and products of its elements. Here the product is operator composition:

$$(T_1 \circ T_2)(x) = T_1 T_2(x) = T_1[T_2 x].$$

$E[X]$ is a semi-group under this operation and all transformation semi-groups acting in X are subsemi-groups of $E[X]$.

Among these the most interesting are those which are *indexed* or *parametrized.* Here the elements of S are of the form $T = T(a)$ where $T(a) \in E[X]$ for each a. Further, a belongs to an index set A which itself is a semi-group under an operation $\circ$ and the relation between composition in A and composition in S is

$$T(a \circ b) = T(a)\, T(b).$$

The simplest instance is that of a *one-parameter transformation semi-group* in which $A = E^+$, the set of positive numbers, and $\circ$ is $+$ so that the elements satisfy the law

$$T(a + b) = T(a)\, T(b) = T(b)\, T(a), \qquad a > 0, \quad b > 0.$$

This case has many important applications and will be discussed further in Section 3 below.

2. AN EXAMPLE

To illustrate these various concepts, we shall consider a simple case involving two-by-two matrices of real numbers

$$a = \begin{pmatrix} a_{11} & a_{12} \\ a_{21} & a_{21} \end{pmatrix}$$

where the operation $\circ$ is matrix multiplication:

$$a \circ b = \begin{pmatrix} a_{11}b_{11} + a_{12}b_{21} & a_{11}b_{12} + a_{12}b_{22} \\ a_{21}b_{11} + a_{22}b_{21} & a_{21}b_{12} + a_{22}b_{22} \end{pmatrix}.$$

The set of all such matrices would form a semi-group under matrix multiplication, but we shall restrict ourselves to nonnegative matrices. Here we say that a is *nonnegative*, $a > 0$, if $a_{jk} \geq 0$. It is clear that the products (and sums) of nonnegative matrices are again nonnegative. Thus the set M^+ of all nonnegative two-by-two positive matrices forms a semi-group under matrix multiplication. M^+ contains the unit matrix e which acts as a unit element in M^+. It should be noted that M^+ is not a group. It contains singular matrices as well as regular ones, but from the fact that a nonnegative matrix has an inverse it does not follow that the inverse is also a nonnegative matrix. This holds only for diagonal matrices if det $(a) > 0$.

There are infinitely many one-parameter subsemi-groups to be found in M^+. Suppose that $p \in M^+$ and form the set

$$M^+(p) \equiv \{\exp(\rho p) \mid 0 \leq \rho < \infty\}.$$

We call such a set an *orbit.* It is a semi-group and all its elements are in M^+. If p is a diagonal matrix, then the group

$$M(p) \equiv \{\exp(\rho p) \mid -\infty < \rho < \infty\}$$

is in M^+, but normally $M^+(p)$ cannot be embedded in a group all the elements of which belong to M^+.

We can introduce a topology in M^+ with the aid of the correspondence

$$(x_1, x_2, x_3, x_4) \longleftrightarrow \begin{pmatrix} x_1 & x_2 \\ x_3 & x_4 \end{pmatrix}$$

between the points of euclidean four-space E_4 on one hand and the two-by-two matrices on the other. We impose the euclidean topology of E_4 on M^+. In this manner we make the algebraic semi-group M^+ into a topological one since matrix multiplication is obviously continuous in the two factors in the topology of E_4.

For all possible choices of p in M^+, the set of orbits forms a system of curves joining the unit matrix e with those elements of M^+ which are either diagonal matrices or have positive logarithms. It is clear that only regular matrices can be on an orbit. But there are also regular positive matrices which do not lie on any orbit. An example is given by

$$\begin{pmatrix} 0 & 1 \\ 1 & 0 \end{pmatrix}$$

which does not admit of any real matrix as its logarithm. We recall that the exponential function of a matrix is defined by the exponential series:

$$\exp p = e + \sum_{n=1}^{\infty} \frac{p^n}{n!},$$

which converges for every matrix p. Further, a matrix y is a logarithm of the matrix x if

$$\exp y = x.$$

We have considered M^+ as an algebraic and as a topological semi-group. There are many ways of associating transformation semi-groups with M^+. We shall give a couple of examples. Matrices are normally used to act either on other matrices or on vectors. We use this observation as a guide in the construction.

The set of all two-by-one matrices with the metric of E_2 is a real Banach space M. Denote temporarily elements of M by x and elements of M^+ by a. Then

$$T(a)x = ax$$

defines a linear bounded transformation on M into itself if a is fixed. If we let a range over M^+, we obtain a transformation semi-group with M^+ as index set. By restricting a to a one-parameter semi-group of M^+, we obtain a one-parameter transformation semi-group as, for instance,

$$\{\exp(\rho p) \mid 0 \leq \rho < \infty, p \in M^+\}.$$

Instead of letting the elements of M^+ act on two-vectors, we can let them act on continuous functions of such vectors, in other words on continuous functions of two real variables. Let E_2^+ denote the first quadrant of the (x, y)-plane and let $C[E_2^+]$ be the set of functions $F(x, y)$ continuous and bounded on E_2^+. This set becomes a Banach space in the usual sup-norm:

$$||F|| = \sup_{(x,y) \in E_2^+} |F(x, y)|.$$

If $a = \begin{pmatrix} a_{11} & a_{12} \\ a_{21} & a_{22} \end{pmatrix} \in M^+$, we define

$$T(a)[F](x, y) = F(a_{11}x + a_{12}y, a_{21}x + a_{22}y).$$

This is obviously a linear bounded transformation on $C[E_2^+]$ to itself, in fact $||T(a)|| \equiv 1$ as we see by taking $F(x, y) \equiv 1$. Here

$$\{T(a) \mid a \in M^+\}$$

is a transformation semi-group with M^+ as index set. We shall not pursue this special case any further.

3. ONE PARAMETER TRANSFORMATION SEMI-GROUPS

One encounters one-parameter transformation semi-groups in a variety of problems of analysis. We shall give some instances.

1. *Translations*: The simplest of all such semi-groups is that of translations. We take $X = C[0, \infty]$ and define

$$T(a)[f](u) = f(u + a), \qquad a \geq 0.$$

The semi-group property is immediate.

2. *Fractional integration*: Let $X = C[0, 1]$ and let

$$T(a)[f](u) = \frac{1}{\Gamma(a)} \int_0^u (u - t)^{a-1} f(t)\, dt, \qquad a > 0,$$

be the fractional integral of order a. The semi-group property has long been known. It holds also for complex values of a with $\Re(a) > 0$. This is the Riemann-Liouville integral. It defines a semi-group of linear bounded transformations on $C[0, 1]$ and on $L_p(0, 1)$. The boundedness is lost on infinite intervals. Marcel Riesz has introduced related integrals for the case of Laplacian, Lorentzian, and other differential operators. These integrals normally define unbounded semi-groups.

3. *Harmonic functions*: If $f(\theta) \in L(-\pi, \pi)$ and $0 < r < 1$, then Poisson's integral

$$f(\theta; r) = \frac{1}{2\pi} \int_0^{2\pi} \frac{(1 - r^2) f(\theta + \phi)\, d\phi}{1 - 2r\cos\phi + r^2}$$

represents the function which is harmonic in the unit circle and takes on the boundary values $f(\theta)$ as $r \longrightarrow 1$. It is also a linear bounded transformation on $L(-\pi, \pi)$ to itself and, if we set $r = e^{-s}$, $s > 0$, and note the expansion

$$f(\theta; e^{-s}) = \sum_{-\infty}^{\infty} f_n e^{ni\theta} e^{-|n| s}, \qquad f(\theta) \sim \sum_{-\infty}^{\infty} f_n e^{ni\theta},$$

we see that these transformations form a one-parameter transformation semi-group. A similar situation holds for a half-plane. The integral

$$T(y)[f](x) = \frac{y}{\pi} \int_{-\infty}^{\infty} \frac{f(u + x)}{u^2 + y^2} du$$

gives the value at the point $x + iy$, $y > 0$, of a function harmonic in the upper half-plane whose boundary values on the real axis are $f(x)$. From this interpretation of the integral the semi-group property becomes almost self-evident. A half-space in three dimensions leads to similar results.

4. *Stochastic processes*: Without giving a specific instance, we just mention that the so-called Chapman-Kolmogoroff equations imply semi-group properties for the corresponding transition probabilities.

5. *Diffusion equations*: This is one of the richest sources of transformation semi-groups that have been found. A couple of elementary instances are given by heat conduction in a circular wire and in an infinite rod. The temperature $T(x, t; f)$ at the place x, after the time t, and initial temperature $f(x)$, is given by

$$\sum_{-\infty}^{\infty} f_n e^{nix - n^2 t}, \qquad f(x) \sim \sum_{-\infty}^{\infty} f_n e^{nix}$$

in the first case, and by

$$(\pi t)^{-1/2} \int_{-\infty}^{\infty} \exp\left(-\frac{u^2}{4t}\right) f(x + u)\, du$$

in the second. The length of the wire is taken to be 2π and the physical constant in the heat equation has been assigned the value 1. These expressions define linear bounded transformations on the initial values, which may be assumed to belong to a Lebesgue space $L_p(-\pi, \pi)$ or $L_p(-\infty, \infty)$. It is easily seen that in both cases we are dealing with one-parameter transformation semi-groups indexed with respect to the time t.

Applications of semi-group methods to the solution of partial differential equations of the parabolic type have been given by several writers including W. Feller, K. Yosida, and the present author. For the hyperbolic case, work by R. S. Phillips should be noted. All these investigations may be considered as specific instances of an abstract Cauchy problem.

6. *Ergodic theory*: Here we are concerned with the infinitary behavior of a semi-group operator. There are two cases according as we consider $\{T^n \mid n = 0, 1, 2, \cdots\}$ or $\{T(s) \mid 0 \leq s\}$. In the discrete case one normally considers the existence of the limit of

$$n^{-1} \sum_{k=0}^{n} T^k,$$

while in the continuous case

$$t^{-1} \int_0^t T(s)\, ds$$

is considered instead. In each case the limit may be taken in the weak or the strong or the uniform sense.

Instead of adding more examples, let us end this discussion by giving a thumbnail sketch of the elements of the theory of one-parameter transformation semi-groups. Thus, we have a Banach space X over complex numbers and a family $\{T(s) \mid 0 < s\}$ of linear bounded transformations on X to itself satisfying the semi-group condition

$$T(s_1 + s_2) = T(s_1)T(s_2), 0 < s_1, s_2 < \infty.$$

$T(s)$ is an operator-valued function of positive numbers. This function $T(s)$ may be quite pathological in its dependence on s, but if we suppose that $T(s)$ is measurable in the strong (in the uniform) operator topology, then $T(s)$ may be shown to be continuous in the strong (in the uniform) sense for $s > 0$.

The norm of $T(s)$ may become infinite as $s \longrightarrow 0$ or to ∞. For large s we have

$$||T(s)|| \leq e^{bs},$$

where b is a fixed finite number, but if $s \longrightarrow 0$, then $||T(s)||$ is not subject to any limitation. For the applications the interesting case is that in which $||T(s)||$ stays bounded and $T(s)$ tends to a limit as $s \longrightarrow 0$. We may assume that this limit is I, the identical operator. There are two essentially different cases according as $T(s) \longrightarrow I$ in the uniform or in the strong sense. We take $b = 0$.

If $$\|T(s) - I\| \longrightarrow 0 \quad \text{as } s \longrightarrow 0,$$
then there exists a linear bounded transformation A such that
$$T(s) = \exp(sA),$$
where the exponential function is defined by the usual series. In this case we can embed the semi-group $\{T(s) \mid 0 < s\}$ in an analytical group $\{\exp(sA)\}$, where s ranges over the complex s-plane.

The second case is much more interesting. Here we have
$$\lim_{s \to 0} \|T(s)x - x\| = 0$$
for each $x \in X$. We can now find a linear *unbounded* operator A whose domain $D(A)$ is dense in X such that
$$\lim_{\delta \to 0} \frac{1}{\delta}[T(\delta)x - x] \equiv Ax$$
exists for each x in $D(A)$. The spectrum of this operator lies in the left half of the λ-plane and for $\Re(\lambda) > 0$, its resolvent $R(\lambda; A)$ is the Laplace transform (in the strong sense) of $T(s)$, that is,
$$R(\lambda; A)x = \int_0^\infty e^{-\lambda s} T(s)x \, ds.$$
Conversely, if $R(\lambda; A)$ is known, we can obtain $T(s)$ by various inversion formulas. Thus, for each $x \in D(A)$ we have
$$T(s)x = \lim_{\omega \to \infty} \frac{1}{2\pi i} \int_{\gamma - i\omega}^{\gamma + i\omega} e^{\lambda s} R(\lambda; A)x \, d\lambda \qquad (\gamma > 0).$$
With the limit replaced by the $(C, 1)$-limit, the formula holds for any x in X. The following formulas are more interesting. We have
$$T(s)x = \lim_{n \to \infty} \left[\frac{n}{s} R\left(\frac{n}{s}; A\right)\right]^n x,$$
$$T(s)x = \lim_{\omega \to \infty} \exp\{s[\omega^2 R(\omega; A) - \omega I]\}x.$$

A is known as the *infinitesimal generator* of the semi-group $T(s)$. Necessary and sufficient conditions are known in order that a transformation A shall generate a semi-group with specified continuity properties. The simplest of these is the so-called Hille-Yosida theorem:

In order that A shall generate a semi-group $T(s)$, such that $T(s) \longrightarrow x$ as $s \longrightarrow 0$ for all x and $||T(s)|| \leq 1$ for $s > 0$, it is necessary and sufficient that A be a closed linear operator whose domain is dense in X and whose resolvent $R(\lambda; A)$ exists for $\lambda > 0$ and satisfies

$$\lambda ||R(\lambda; A)|| \leq 1.$$

In the second, strong case, $T(s)x$ is a continuous function of s for $s \geq 0$. If $x \in D(A)$, then $T(s)x$ is also differentiable in the strong sense and

$$\frac{d}{ds} T(s)x = AT(s)x = T(s)Ax.$$

This relation is basic for the abstract Cauchy problem which calls for a solution of the functional equation

$$\frac{dU}{ds} = AU$$

which belongs to X for $s > 0$ and tends to a preassigned limit x in X as $s \longrightarrow 0$. If A satisfies the conditions of the Hille-Yosida theorem, and if $x \in D(A)$, then the problem has a unique solution given by $T(s)x$.

$T(s)x$ has a derivative of order k with respect to s if x belongs to the domain of A^k which is also dense in X. In particular, $T(s)$ has derivatives of all orders for $s > 0$ if $AT(s)$ is a bounded operator for each $s > 0$. This condition implies that $T(s)$ maps all of X into $D(A)$. From the fact that such an operator $T(s)$ has derivatives of all orders, it does not follow that $T(s)$ is analytic. There exist analytic semi-group operators, however; thus the operators under instances 2, 3, and 5 above are analytic in the right half-plane. This is the maximum domain of analyticity; in general the domain of analyticity of a semi-group operator is a simply connected domain which is a semi-group under addition. It is possible to give conditions for analyticity of $T(s)$ in terms of properties of the resolvent of the infinitesimal generator A. The main condition is that the spectrum of A must recede from the imaginary axis in the λ-plane. In other words, if $\{\lambda_n\}$ belongs to the spectrum of A and if $|\lambda_n| \longrightarrow \infty$, then $\Re(\lambda_n) \longrightarrow -\infty$.

With this we end our survey and refer the interested reader to the books and articles listed below.

REFERENCES

1. Dunford, N., and J. T. Schwartz, *Linear Operators.* New York: Interscience Publishers, 1958, pt. 1, chap. 8.
2. Feller, W., "The parabolic differential equation and the associated semi-group of transformations," *Ann. Math.*, (2) vol. 55 (1952), pp. 468–519.
3. Hille, E., "The abstract Cauchy problem and Cauchy's problem for parabolic differential equations," *J. Analyse Math.*, vol. 3 (1953), pp. 81–196, 414.
4. Hille, E., and R. S. Phillips, "Functional analysis and semi-groups," *Coll. Lect. Amer. Math. Soc.*, vol. 31, 2nd ed. Providence, R.I. (1957).
5. Neveu, J., *Théorie des semi-groupes de Markov.* Berkeley and Los Angeles: University of California Press, 1958.
6. Phillips, R. S., "Dissipative hyperbolic systems," *Trans. Amer. Math. Soc.*, vol. 86 (1957), pp. 109–173.
7. Schwartz, L., *Lectures on Mixed Problems in Partial Differential Equations and Representations of Semi-Groups.* Bombay: Tata Institute of Fundamental Research, 1958.
8. Wallace, A. D., "The structure of topological semi-groups," *Bull. Amer. Math. Soc.*, vol. 61 (1955), pp. 95–112.
9. Yosida, K., *Lectures on Semi-Group Theory and Its Application to Cauchy's Problem in Partial Differential Equations.* Bombay: Tata Institute of Fundamental Research, 1957.

THE LAPLACE TRANSFORM, THE STIELTJES TRANSFORM, AND THEIR GENERALIZATIONS

I. I. Hirschman, Jr., and D. V. Widder

1. INTRODUCTION

We begin with a study of the Laplace transform, which is an integral of the form

$$f(s) = \int_0^\infty e^{-st}\phi(t)\,dt, \tag{1}$$

where $\phi(t)$ is any function which, for some value of s, gives the integral meaning. The integral then exists for a whole interval of values of s, so that a function $f(s)$ is defined. Since equation (1) may be thought of as *transforming* $\phi(t)$ into $f(s)$, it is frequently called the *Laplace transform*.

If, for example, $\phi(t) = 1$, then

$$f(s) = \lim_{R\to\infty} \int_0^R e^{-st}\,dt = \lim_{R\to\infty} \frac{1 - e^{-sR}}{s}, \qquad s \neq 0. \tag{2}$$

When $s > 0$ this limit is $1/s$; when $s \leq 0$ there is no limit. More generally, if s is complex, $s = \sigma + i\tau$, the limit (2) exists or fails to exist according as σ is positive or negative. The general integral (1) behaves similarly: it *converges* in a right half-plane, $\sigma > \sigma_c$, and *diverges* for $\sigma < \sigma_c$. The number σ_c, which may be $+\infty$ or $-\infty$, is called the *abscissa of convergence.*

The name for the integral (1) was chosen because Laplace† used it extensively in his theory of probability. The modern revival of interest in the transform probably was caused by Riemann's discovery that the distribution of prime numbers depends upon the position of the zeros of the zeta-function,

$$\zeta(s) = \sum_{n=1}^{\infty} \frac{1}{n^s}. \tag{3}$$

His famous conjecture, still unverified, that any zero of this function with positive real part must have the imaginary part 1/2 has had a tremendous influence on the development of mathematics. Now $\zeta(s)/s$ is a Laplace integral. Accordingly it was very natural that in looking for specific properties of the special function (3) the general properties of the integral (1) should have been discovered.

In this brief note no attempt will be made to give references for the various results described. They may be found in the treatises listed in the bibliography.

2. RELATION TO POWER SERIES

A natural way of generalizing a power series

$$F(z) = \sum_{n=0}^{\infty} a_n z^n \tag{4}$$

is to replace the integral exponent n by an arbitrary real number λ_n. But for complex z the function z^{λ_n} usually has many values. One convenient way of specifying which value is intended is to set $z = e^{-s}$,

---

† See, for example, pp. 111 to 180 in vol. VII of the collected works of Laplace.

$$F(e^{-s}) = \sum_{n=0}^{\infty} a_n e^{-\lambda_n s}. \tag{5}$$

If λ_n tends to $+\infty$, this is a *Dirichlet series*. The zeta-function is seen to be a special case of the series (5) by taking $\lambda_n = \log n$.

It is now natural to generalize a step further by changing λ_n to a continuous variable t. The summation becomes an integral, the sequence a_n becomes a function $\phi(t)$, and the Dirichlet series becomes the Laplace integral. Accordingly it is not surprising that many of the properties of the integral can be correctly conjectured from the corresponding ones for power series. For example, the region of convergence of a power series is $|z| < \rho$. Hence we might expect that the region of convergence of a Laplace integral is $|e^{-s}| = e^{-\sigma} < \rho$ or $\sigma > -\log \rho$, a right half-plane. We have stated this fact and verified it in a particular example.

Let us list several properties that do carry over from power series:

	Power series	*Laplace integrals*
1. Convergence	$\lvert z\rvert < \rho$	$\sigma > \sigma_c$
2. Differentiation	$F'(z) = \sum_{n=0}^{\infty} n a_n z^{n-1}$	$f'(s) = -\int_0^{\infty} e^{-st} t\, \phi(t)\, dt$
3. Analyticity	$\lvert z\rvert < \rho$	$\sigma > \sigma_c$
4. Uniqueness	$F \equiv 0$ implies $a_n \equiv 0$	$f \equiv 0$ implies $\phi(t) \equiv 0$
5. Inversion	$a_n = \frac{1}{2\pi i}\int_{\lvert z\rvert=k} F(z) z^{-n-1}\, dz$	$\phi(t) = \frac{1}{2\pi i}\int_{c-i\infty}^{c+i\infty} f(s) e^{st}\, ds$
6. Products	$FG = \sum_{n=0}^{\infty} c_n z^n$	$fg = \int_0^{\infty} e^{-st}\omega(t)\, dt$
	$c_n = \sum_{k=0}^{n} a_k b_{n-k}$	$\omega(t) = \int_0^t \phi(x)\psi(t-x)\, dx$

In property 4 it must be understood that $\phi(t)$ is *essentially* zero. It may be different from zero at some points, but at any rate its integral over any interval must vanish. In 5, $k < \rho$ and $c > \sigma_c$. Here the analogy is not quite complete since the substitution $z = e^{-s}$ carries the circle $|z| = k$ into only a piece of the vertical line $\sigma = c = -\log k$. In 6, the power series expansion of $G(z)$ has coefficients b_n and $g(s)$ is the Laplace transform of $\psi(t)$.

It is also important to observe certain fundamental differences between the two theories. Whereas a power series converges absolutely inside its region of convergence the same is not true for Laplace integrals. Indeed for the latter there are abscissas of conditional, absolute, and uniform convergence, generally all different. Series (4) converges out to the singularity of $F(z)$ nearest the origin; $f(s)$ need have no singularity on the line $\sigma = \sigma_c$. Again, every analytic function has a power series expansion; the function $f(s) = s$ is entire but is not the Laplace transform of any function. Finally, there is invariably a difference in the methods of proof in the two theories. Often the gap is bridged by an integration by parts, for this process replaces a conditionally convergent Laplace integral by one which converges absolutely.

3. BILATERAL LAPLACE TRANSFORM

We shall later have occasion to use the bilateral Laplace transform, which is obtained by replacing the lower limit of integration in the integral (1) by $-\infty$. The bilateral Laplace transform is analogous to a Laurent series and its region of convergence is a vertical strip. For future reference we compute the bilateral Laplace transform of $e^t e^{-e^t}$; it is

$$\int_{-\infty}^{\infty} e^{-st} e^t e^{-e^t}\, dt = \int_0^{\infty} x^{-s} e^{-x}\, dx = \Gamma(1 - s)$$

for $-\infty < \operatorname{Re} s < 1$. In this case, the vertical strip has been enlarged into a half-plane.

4. COMPLEX INVERSION FORMULA

We shall need to use the complex inversion formula of the bilateral Laplace transform. If the region of absolute convergence of the bilateral Laplace transform

$$\int_{-\infty}^{\infty} \phi(t) e^{-st}\, dt = \Phi(s)$$

includes the line Re $s = c$, then

$$\phi(t) = \frac{1}{2\pi i} \lim_{T\to\infty} \int_{c-iT}^{c+iT} \Phi(s)e^{st}\, ds$$

for all values of t at which $\phi(t)$ is smooth enough. (For example, differentiability at a point is sufficient.) In what follows, the bilateral Laplace transform (equivalent to the Fourier transform after a simple change of variables) and the Laplace transform play quite different roles. The Laplace transform will ultimately appear as a special example of a large class of integral transforms. In studying this class of integral transforms, the bilateral Laplace transform enters as a basic tool. It is important to keep in mind the very dissimilar roles played by these two transformations which seem at first glance to differ so slightly.

5. A TABLE OF TRANSFORMS

It is useful to have a table of Laplace transforms to be used like any table of integrals. Here is a highly abridged one:

	$f(s)$	$\phi(t)$	σ_c
A.	$\frac{1}{s}$	1	0
B.	$\frac{1}{s^{n+1}}$	$\frac{t^n}{n!}$	0
C.	$\frac{1}{s-a}$	e^{at}	α
D.	$\frac{1}{s^2+1}$	$\sin t$	0
E.	$\frac{s}{s^2+1}$	$\cos t$	0

The pair B may be derived from A by differentiation; C comes from the same source when s is replaced by $s - a$. Here α is the real part of a. Both D and E come from C when $\sin t$ and $\cos t$ are expressed in terms of e^{it} and e^{-it}.

6. DIFFERENTIAL EQUATIONS

As an illustration of the many applications of the general theory let us solve a differential equation. It will be clear from the example that the method is very general. The transform (1) applied to an ordinary linear differential equation with constant coefficients reduces it to an algebraic one. The solution of the latter is then retransformed by inversion, or by use of a table. More generally, if the original equation is partial in any number of independent variables, one application of the Laplace transform reduces the number of variables by one.

Let it be required to find a function $y(t)$ such that $y(0) = 1$, $y'(0) = 2$, and

$$y'' + y = 2e^t. \tag{6}$$

Denote the Laplace transform of $y(t)$ by $Y(s)$. Integration by parts gives

$$\int_0^\infty e^{-st} y''(t)\, dt = -y'(0) - y(0)s + s^2 \int_0^\infty e^{-st} y(t)\, dt$$

on the assumption that the integrated part vanishes at $+\infty$, at least for large values of s. Using the pair C of Sec. 5 we see that the transformed equation is

$$-2 - s + s^2 Y(s) + Y(s) = \frac{2}{s-1}$$

$$Y(s) = \frac{s^2 + s}{(s-1)(s^2+1)} = \frac{1}{s-1} + \frac{1}{s^2+1}.$$

A second reference to the table and an appeal to the uniqueness property 4 shows that $y(t) = e^t + \sin t$. This function has the required properties, so that it is unnecessary to check the assumption made about its behavior at $+\infty$.

The procedure is elementary and could be used in an introductory study of differential equations. The proof that the method always leads to the solution has hitherto depended on contour integration in the complex plane. In a text on advanced calculus by the second author a simpler proof depending only on real variable theory is now available.

7. REAL INVERSION

The most familiar determination of the coefficients of the power series (4) is $a_n = F^{(n)}(0)/n!$. An analogous inversion of the Laplace transform (1) has recently been discovered. It is

$$\phi(t) = \lim_{k\to\infty} \frac{(-1)^k}{k!} f^{(k)}\left(\frac{k}{t}\right)\left(\frac{k}{t}\right)^{k+1}. \tag{7}$$

Observe that $\phi(t)$ is determined by the values of the successive derivatives of $f(s)$ near $s = +\infty$. Since $z = 0$ corresponds to $s = +\infty$ when $z = e^{-s}$, formula (7) is indeed similar to the Taylor "inversion" of a power series. For the pair $f(s) = s^{-2}$, $\phi(t) = t$ Eq. (7) becomes

$$t = \lim_{k\to\infty}\left(1 + \frac{1}{k}\right)t.$$

For the pair C it is

$$e^{at} = \lim_{k\to\infty}\left(1 - \frac{at}{k}\right)^{-k-1}.$$

8. RELATION TO THE MOMENT PROBLEM

The *moment problem* of Hausdorff is the determination of a nondecreasing function $\beta(x)$ such that

$$\mu_n = \int_0^1 x^n\, d\beta(x), \qquad n = 0, 1, 2, \cdots, \tag{8}$$

where the integral is a Stieltjes integral. Hausdorff showed that the problem has a solution if and only if, for all n,

$$\mu_n \geq 0,\ \Delta\mu_n = \mu_{n+1} - \mu_n \leq 0,\ \Delta^2\mu_n = \mu_{n+2} - 2\mu_{n+1} + \mu_n \geq 0,\ \cdots.$$

Such a sequence is given by $\mu_n = 1/(n + 1)$ or by $(1/2)^n$, and these arise from Eq. (8) by taking first $\beta(x) = x$ and then $\beta(x) = 0$ or 1 according as $x < 1/2$ or $x > 1/2$. Both functions are nondecreasing.

If in Eq. (8) we replace n by a continuous variable s and set $x = e^{-t}$ we obtain

$$\mu(s) = \int_0^\infty e^{-st}\, d[-\beta(e^{-t})], \qquad 0 \le s < \infty.$$

But this is a *Laplace-Stieltjes transform*

$$f(s) = \int_0^\infty e^{-st}\, d\alpha(t). \tag{9}$$

By analogy with the Hausdorff result we might expect that a nondecreasing function $\alpha(t)$ would exist satisfying Eq. (9) if and only if

$$f(s) \ge 0, \quad f'(s) \le 0, \quad f''(s) \ge 0, \quad \cdots, \qquad 0 < s < \infty.$$

Such a function is called *completely monotonic.* Examples are $f(s) = 1/(s+1)$ and $f(s) = e^{-s}$ and these arise from Eq. (9) by use of the nondecreasing functions $\alpha(t) = -e^{-t}$ and $\alpha(t) = 0$ or 1 according as $t < 1$ or $t > 1$. The conjectured result is true and is known as Bernstein's theorem. The result is particularly remarkable in view of the fact that the signs of the derivatives of a function on the real axis should determine not only its analyticity in a half-plane but its representation in the form (9).

At first sight it may not be easy to see why the Stieltjes integral has been introduced at this stage. It is done largely to produce elegant results like those of Hausdorff and Bernstein given above. The function e^{-s}, though completely monotonic, could never have a representation (1). The same is true of any convergent Dirichlet series with positive coefficients. It is only when we introduce the Stieltjes integral, thus combining the class of such Dirichlet series with the functions defined by Eq. (1) with $\phi(t) \ge 0$, that we can obtain a neat characterization of completely monotonic functions.

9. THE STIELTJES TRANSFORM

If the transform (1) is applied to itself there results, after a change in the order of integration,

$$f(s) = \int_0^\infty e^{-sx}\, dx \int_0^\infty e^{-xt}\phi(t)\, dt = \int_0^\infty \frac{\phi(t)}{s+t}\, dt. \tag{10}$$

This is called the *Stieltjes transform* because it was put to effective use by Stieltjes in his theory of continued fractions. A sample pair is $f(s) = \pi/\sqrt{s}$, $\phi(t) = 1/\sqrt{t}$. This may be verified by the change of variable $t = u^2$. From the origin of the transform as an *iterated* Laplace transform many of its properties can be conjectured. An inversion formula related to property 5 of Sec. 2 is

$$\phi(t) = \lim_{\epsilon \to 0} \frac{f(-t - i\epsilon) - f(-t + i\epsilon)}{2\pi i}. \tag{11}$$

Another, matching the one given in Sec. 7, is

$$\phi(t) = \lim_{k \to \infty} \frac{(-t)^{k-1}}{k!(k-2)!} [t^k f(t)]^{(2k-1)}. \tag{12}$$

It is interesting to test this equation with the pair of transforms given above.

A result analogous to Bernstein's is that a nonnegative number P and a nondecreasing function $\alpha(t)$ such that

$$f(s) = P + \int_0^\infty \frac{d\alpha(t)}{s + t}$$

exist if and only if

$$f(s) \geq 0, \quad (-1)^{k-1}[s^k f(s)]^{(2k-1)} \geq 0, \qquad 0 < s < \infty, k = 1, 2, \cdots.$$

The function $f(s) = \pi/\sqrt{s}$ has this property, and the corresponding function $\alpha(t) = 2\sqrt{t}$ is clearly nondecreasing.

10. THE OPERATIONAL CALCULUS

Let L_c^1 (L_p^1) be the class of continuous (piecewise continuous) functions $g(x)$ defined for $-\infty < x < \infty$ and such that the integral

$$||g||_1 = \int_{-\infty}^\infty |g(x)| \, dx$$

is finite. If $g(t)$ and $h(t)$ belong to L_c^1, then their (bilateral) convolution $f(x)$ is defined by

$$f(x) = \int_{-\infty}^\infty g(x - t)h(t) \, dt, \qquad -\infty < x < \infty. \tag{13}$$

We express this symbolically as $f = g * h$. A simple change of orders of integration shows that $||g * h||_1 \leq ||g||_1 \, ||h||_1$. Moreover it is easy to see that $g * h(x)$ is a continuous function of x. Making the change of variable $x - t = u$ in (13), we find that $g * h = h * g$; that is, "$*$" is a commutative operation. If g_1, g_2, and g_3 belong to L_c^1, we can in view of the remarks above form both $(g_1 * g_2) * g_3$ and $g_1 * (g_2 * g_3)$. A simple inversion of orders of integration and change of variables shows that these two triple convolutions are equal; that is, "$*$" is associative.

If $g(t)$ is in L_c^1, its bilateral Laplace transform defined in Sec. 3,

$$\hat{g}(s) = \int_{-\infty}^{\infty} g(t)e^{-st}\, dt, \tag{14}$$

converges at least on the line $\operatorname{Re} s = 0$ and possibly, depending upon g, in a vertical strip including this line. Let $f = g * h$, where g and h belong to L_c^1; then

$$\hat{f}(s) = \int_{-\infty}^{\infty} e^{-st}\, dt \int_{-\infty}^{\infty} g(t-u)h(u)\, du$$

$$\hat{f}(s) = \int_{-\infty}^{\infty} h(u)e^{-su}\, du \int_{-\infty}^{\infty} e^{-s(t-u)} g(t-u)\, dt$$

so that at least for $\operatorname{Re} s = 0$ we have the striking formula

$$(g * h)\hat{\ }(s) = \hat{g}(s)\hat{h}(s). \tag{15}$$

See property 6 of Sec. 2.

It is evident from these considerations that convolution is a very natural way of combining functions. It is thus significant that both the (one-sided) Laplace transform and the Stieltjes transform can, after a change of variables, be written in this form. If Eq. (1) is transformed by replacing s by e^x and t by e^{-u}, it becomes

$$f(e^x)e^x = \int_{-\infty}^{\infty} e^{x-u} \exp(-e^{x-u})\phi(e^{-u})\, du. \tag{16}$$

In a similar way (10) becomes

$$f(x)e^{x/2} = \frac{1}{2}\int_{-\infty}^{\infty} \operatorname{sech} \frac{x-u}{2}\, e^{u/2}\phi(e^{u/2})\, du. \tag{17}$$

When $G(x)$ is a fixed function in L_p^1 we call the integral transformation

$$h(x) = \int_{-\infty}^{\infty} G(x-u)\psi(u)\,du, \tag{18}$$

which sends ψ in L_c^1 into h in L_c^1, the convolution transform with kernel G. Thus both the Laplace and Stieltjes transforms appear after a change of variables as convolution transforms with kernels $e^x e^{-e^x}$ and $\frac{1}{2}$ sech $(x/2)$, respectively. It is natural to inquire if the formulas (16) and (17) are not in some way connected with the inversion formulas (7) and (12). We shall show that this is indeed the case. Let the symbol D stand for differentiation with respect to x. Since the symbolic series

$$e^{aD}h(x) = \sum_{k=0}^{\infty} \frac{a^k}{k!} D^k h(x) = \sum_{k=0}^{\infty} \frac{a^k}{k!} h^{(k)}(x)$$

is equivalent to Taylor's series, whose sum under favorable circumstances is $h(x+a)$, we *define*

$$e^{aD}h(x) = h(x+a).$$

Now let us examine the result of applying the bilateral Laplace transform to $Df(x)$ and to $e^{aD}f(x)$. Suppose first $f(x)$ and $f'(x)$ belong to L_c^1. Then

$$\begin{aligned}(Df)^{\wedge}(s) &= \int_{-\infty}^{\infty} e^{-st}f'(t)\,dt \\ &= [e^{-st}f(t)]_{-\infty}^{\infty} + s\int_{-\infty}^{\infty} e^{-st}f(t)\,dt \\ &= s\hat{f}(s).\end{aligned}$$

More generally, if $P(s) = a_0 + a_1 s + \cdots + a_n s^n$ is a polynomial of degree n and if $f, f^1, \cdots, f^{(n)}$ all belong to L_c^1, then

$$[P(D)f]^{\wedge}(s) = P(s)\hat{f}(s), \qquad \text{Re}\, s = 0. \tag{19}$$

Again if f is in L_c^1

$$(e^{aD}f)^{\wedge}(s) = \int_{-\infty}^{\infty} e^{-st}f(t+a)\,dt = e^{as}\hat{f}(s). \tag{20}$$

Finally we will need the delta function of Dirac. This is a symbol $\delta(x)$ which is defined to behave as though it were a function on $-\infty < x < \infty$, equal to 0 for all x except $x = 0$ where it is equal to $+\infty$. The $+\infty$ is of such a nature that $\int_{-\infty}^{\infty} \delta(x)\,dx = 1$. Essentially we are defining δ to have the property that if $f(x)$ is any function defined on $-\infty < x < \infty$ and continuous at $x = a$ then

$$\int_{-\infty}^{\infty} \delta(x-a) f(x)\,dx = f(a). \tag{21}$$

In particular

$$\delta^{\wedge}(s) = \int_{-\infty}^{\infty} e^{-st}\delta(t)\,dt = e^{-st}\Big|_{t=0} = 1; \tag{22}$$

that is, the bilateral Laplace transform of δ is the constant 1.

Now consider the bilateral Laplace transform of $e^{x}e^{-e^{x}}$; we find that (see Sec. 3)

$$\int_{-\infty}^{\infty} e^{t}e^{-e^{t}}e^{-st}\,dt = \Gamma(1-s) \tag{23}$$

$$= \frac{1}{e^{-\gamma s}\prod_{k=1}^{\infty}\left(1-\frac{s}{k}\right)e^{s/k}},$$

where γ is Euler's constant

$$\gamma = \lim_{n\to\infty}\left(\sum_{1}^{n} k^{-1} - \log n\right).$$

Using the formulas (19) and (20) we find, entirely rigorously, that if

$$G_n(x) = \left[e^{D\log n}\prod_{k=1}^{n}\left(1-\frac{D}{a_k}\right)\right]e^{x}e^{-e^{x}},$$

then

$$G_n^{\wedge}(s) = \frac{e^{s\log n}\prod_{1}^{n}\left(1-\frac{s}{k}\right)}{e^{-s\gamma}\prod_{1}^{\infty}\left(1-\frac{s}{k}\right)e^{s/k}} \tag{24}$$

$$= \frac{1}{e^{\epsilon_n s}\prod_{n+1}^{\infty}\left(1-\frac{s}{k}\right)e^{s/k}}, \qquad \operatorname{Re} s = 0.$$

Here $\epsilon_n = -\gamma - \log n + \sum_1^n k^{-1}$ so that $\epsilon_n \longrightarrow 0$ as $n \longrightarrow \infty$. It is evident from formula (24) that for Re $s = 0$

$$\lim_{n\to\infty} G_n^\wedge(s) = 1 = \overset{\wedge}{\delta}(s).$$

This suggests that in some sense

$$\lim_{n\to\infty} G_n(x) = \delta(x). \tag{25}$$

Let us proceed as if (25) were true.

If $\psi(x)$ is in L_c^1 and if

$$h(x) = \int_{-\infty}^{\infty} e^{x-u}e^{e^{(x-u)}}\psi(u)\,du$$

then it is easy to see (property 2 of Sec. 2) that we may differentiate under the integral sign as often as we please and in particular that

$$e^{D\log n}\prod_1^n\left(1-\frac{D}{k}\right)h(x) = \int_{-\infty}^{\infty} G_n(x-u)\psi(u)\,du. \tag{26}$$

Using the heuristic formula (25) we have formally that

$$\lim_{n\to\infty} e^{D\log n}\prod_1^n\left(1-\frac{D}{k}\right)h(x) = \int_{-\infty}^{\infty} \delta(x-u)\psi(u)\,du = \psi(x). \tag{27}$$

Let us refer back to (16); we see that if

$$f(x) = \int_0^{\infty} e^{-xt}\psi(t)\,dt$$

then (formally)

$$\lim_{n\to\infty} e^{D\log n}\prod_1^n\left(1-\frac{D}{k}\right)e^x f(e^x) = \psi(e^{-x}). \tag{28}$$

A simple computation shows that

$$(1-D)e^x f(e^x) = -e^{2x}f'(e^x),$$

$$\left(1-\frac{D}{2}\right)(1-D)e^x f(e^x) = \frac{1}{2!}e^{3x}f''(e^x),$$

$$\vdots$$

$$\prod_1^n \left(1 - \frac{D}{k}\right) e^x f(e^x) = \frac{(-1)^n}{n!} e^{(n+1)x} f^{(n)}(e^x),$$

$$e^{D \log n} \prod_1^n \left(1 - \frac{D}{k}\right) e^x f(e^x) = \frac{(-1)^n}{n!} (ne^x)^{n+1} f^{(n)}(ne^x).$$

Setting $e^{-x} = t$ in (28) we obtain (7),

$$\lim_{n\to\infty} \frac{(-1)^n}{n!} \left(\frac{n}{t}\right)^{n+1} f^{(n)}(t) = \psi(t).$$

Similarly starting from the formula

$$(29) \quad \int_{-\infty}^{\infty} e^{-st} \frac{1}{2} \operatorname{sech} \frac{t}{2}\, dt = \frac{\pi}{\cos \pi s}, \qquad \operatorname{Re} s = 0$$

$$= \frac{\pi}{\prod_1^\infty \left[1 - \frac{s}{\left(k - \frac{1}{2}\right)}\right]\left[1 + \frac{s}{\left(k - \frac{1}{2}\right)}\right]}$$

we can derive (12).

11. THE CONVOLUTION TRANSFORM

The work done in Sec. 10 suggests the following. Let $b, a_1, a_2, \cdots$ be real nonzero constants such that $\Sigma_1^\infty a_k^{-2} < \infty$, and let

$$(30) \qquad E(s) = e^{bs} \prod_1^\infty \left(1 - \frac{s}{a_k}\right) e^{s/a_k}.$$

Suppose there exists a function $G(t)$, which together with all its derivatives belongs to L_c^1, such that

$$(31) \qquad \hat{G}(s) = \frac{1}{E(s)}, \qquad \operatorname{Re} s = 0.$$

Then we may expect the convolution transform with G as kernel

$$h(x) = \int_{-\infty}^{\infty} G(x - u)\psi(u)\, du$$

to have the inversion formula

$$(32) \qquad \lim_{n\to\infty} e^{b_n D} \prod_1^n \left(1 - \frac{D}{a_k}\right) e^{D/a_k} h(x) = \psi(x).$$

Here $b_1, b_2, \cdots$ is any sequence of real numbers such that $\lim_{n\to\infty} b_n = b$. Indeed if we define

$$G_n(x) = e^{b_n D} \prod_1^n \left(1 - \frac{D}{a_k}\right) e^{D/a_k} G(x)$$

then (rigorously)

$$G_n^\wedge(s) = \frac{1}{e^{(b-b_n)s} \prod_{n+1}^{\infty} \left(1 - \frac{s}{a_k}\right) e^{s/a_k}}, \qquad \text{Re } s = 0,$$

and thus

$$\lim_{n\to\infty} G_n^\wedge(s) = 1 = \delta^\wedge(s),$$

which suggests that in some sense

$$\lim_{n\to\infty} G_n(x) = \delta(x). \tag{33}$$

Since

$$e^{b_n D} \prod_1^n \left(1 - \frac{D}{a_k}\right) e^{D/a_k} h(x) = \int_{-\infty}^{\infty} G_n(x-u)\psi(u)\, du$$

we have formally that

$$\lim_{n\to\infty} e^{b_n D} \prod_1^n \left(1 - \frac{D}{a_k}\right) e^{D/a_k} h(x) = \int_{-\infty}^{\infty} \delta(x-u)\psi(u)\, du = \psi(x).$$

The theory sketched above can be completed if we can show that given $E(s)$ there exists a $G(t)$ as described and if we can validate the formal relation (33).

12. ELEMENTARY KERNELS

The detailed solution of the problems raised in Sec. 11 is too long to carry out completely; however, we will briefly indicate the main ideas involved. If $G^\wedge$ is to satisfy (31), then by the complex inversion formula for the bilateral Laplace transform (see Sec. 4) we should have

$$G(t) = \frac{1}{2\pi i} \int_{-i\infty}^{i\infty} [E(s)]^{-1} e^{st}\, ds. \tag{34}$$

This suggests that we take (34) as the *definition* of $G(t)$. It can then in fact be verified that $G(t)$ together with all its derivatives belongs to L_c^1 and that (31) is satisfied.

It remains to make sense out of (33). Let us define

$$g(t) = \begin{cases} e^t, & t < 0, \\ \frac{1}{2}, & t = 0, \\ 0, & t > 0. \end{cases}$$

Clearly $g(t)$ belongs to L_p^1. It is easy to verify that

$$g^\wedge(s) = \frac{1}{1 - s}, \qquad \text{Re } s = 0.$$

Given real constants $a_1, a_2, a_3, \cdots$ such that $\sum_1^\infty a_k^{-2} < \infty$, let

$$g_k(t) = |a_k| g(a_k t), \qquad k = 1, 2, \cdots.$$

It follows from the above that g_k is in L_p^1 and that

$$g_k^\wedge(s) = \frac{1}{\left(1 - \dfrac{s}{a_k}\right)}, \qquad \text{Re } s = 0.$$

The functions $g_k(t)$ will be called elementary kernels. From (15) we see that

$$\int_{-\infty}^{\infty} e^{-st}[g_1 * g_2 * \cdots * g_n(t)]\, dt = \frac{1}{\prod_1^n \left(1 - \dfrac{s}{a_k}\right)}, \qquad \text{Re } s = 0,$$

so that if

$$H_k(t) = g_1 * \cdots * g_n \cdot \left(t - b - \sum_{j=1}^{k} a_j^{-1}\right)$$

then

$$H_k^\wedge(s) = \frac{1}{e^{bs} \prod_1^n \left(1 - \dfrac{s}{a_k}\right) e^{s/a_k}}, \qquad \text{Re } s = 0.$$

The complex inversion formula of Sec. 4 gives

$$H_k(t) = \frac{1}{2\pi i} \int_{-i\infty}^{i\infty} \left[e^{bs} \prod_{j=1}^{k} \left(1 - \frac{s}{a_j}\right) e^{s/a_j} \right]^{-1} e^{st}\, ds.$$

Comparing this with (34) it is easy to show that

$$G(t) = \lim_{k\to\infty} H_k(t), \qquad -\infty < t < \infty. \tag{35}$$

Now each $H_k(t)$ is a finite convolution of nonnegative functions and is therefore nonnegative. It follows from (35) that $G(t)$ is nonnegative. Since $G_n(t)$ is analogous to $G(t)$, with

$$e^{(b-b_n)s} \prod_{n+1}^{\infty} \left(1 - \frac{s}{a_n}\right) e^{s/a_k} \quad \text{in place of} \quad e^{bs} \prod_{1}^{\infty} \left(1 - \frac{s}{a_k}\right) e^{s/a_k},$$

it follows that

$$G_n(t) \geq 0, \qquad -\infty < t < \infty; \quad n = 0, 1, \cdots. \tag{36}$$

We have for Re $s = 0$

$$\int_{-\infty}^{\infty} G_n(t) e^{-st}\, dt = \hat{G_n}(s) = \frac{1}{e^{(b-b_n)s} \prod_{n+1}^{\infty} \left(1 - \frac{s}{a_k}\right) e^{s/a_k}}. \tag{37}$$

Setting $s = 0$ in (37) yields

$$\int_{-\infty}^{\infty} G_n(t)\, dt = 1, \qquad n = 1, 2, \cdots. \tag{38}$$

The formulas (36) and (38) show that the functions $G_n(t)$ and $G(t)$ are probability frequency functions. This suggests computing the mean and variance of $G_n(t)$. We have

$$-\log G_n^{\wedge}(s) = (b - b_n)s + \sum_{k=n+1}^{\infty} \left\{ \log\left(1 - \frac{s}{a_k}\right) + \frac{s}{a_k} \right\}.$$

Differentiating with respect to s and setting $s = 0$ we see that

$$-\frac{G_n^{\wedge\prime}(0)}{G_n^{\wedge}(0)} = b - b_n. \tag{39}$$

Similarly if we differentiate twice with respect to s and then set $s = 0$ we obtain

$$-\frac{G_n^{\wedge}(0) G_n^{\wedge\prime\prime}(0) - G_n^{\wedge\prime}(0)^2}{G_n^{\wedge}(0)^2} = -\sum_{n+1}^{\infty} a_k^{-2}. \tag{40}$$

From (39) and (38) it follows that

$$\int_{-\infty}^{\infty} G_n(t) t\, dt = b - b_n, \tag{41}$$

and from (40) we can show that

$$\int_{-\infty}^{\infty} G_n(t)[t - (b - b_n)]^2\, dt = \sum_{k=n+1}^{\infty} a_k^{-2}. \tag{42}$$

The formulas (37), (38), (41), and (42) may be used as a rigorous equivalent of (33). Indeed, suppose that $\psi(u)$ is continuous and bounded for $-\infty < u < \infty$. Fix x; then given $\epsilon > 0$ we can choose $\delta > 0$ so small that $|\psi(u) - \psi(x)| < \epsilon$ for $|u - x| < \delta$. If

$$h(x) = \int_{-\infty}^{\infty} G(x - t)\psi(u)\, du$$

then, as we have seen,

$$h_n(x) = e^{b_n D} \prod_1^n \left(1 - \frac{D}{a_k}\right) e^{D/a_k} h(x),$$

$$= \int_{-\infty}^{\infty} G_n(x - u)\psi(u)\, du.$$

Consequently, by (38),

$$h_n(x) - \psi(x) = \int_{-\infty}^{\infty} G_n(x - u)[\psi(u) - \psi(x)]\, du,$$

and thus

$$|h_n(x) - \psi(x)| \leq I_1 + I_2,$$

where

$$I_1 = \int_{|u-x| \leq \delta} G_n(x - u)\, |\psi(u) - \psi(x)|\, du,$$

$$I_2 = \int_{|u-x| > \delta} G_n(x - u)\, |\psi(u) - \psi(x)|\, dx.$$

We have

$$I_1 \leq \epsilon \int_{|u-x| \leq \delta} G_n(x - u)\, du \leq \epsilon \int_{-\infty}^{\infty} G_n(x - u)\, du = \epsilon.$$

If $M = \underset{-\infty < u < \infty}{\text{l.u.b.}} |\psi(u)|$ then

$$I_2 \leq 2M \int_{|u-x| > \delta} G_n(x - u)\, du,$$

$$\leq 2M \left(\frac{\delta}{2}\right)^{-2} \int_{-\infty}^{\infty} G_n(x - u)[x - u - (b - b_n)]^2\, du$$

$$= 8M\delta^{-2} \sum_{k=n+1}^{\infty} a_k^{-2},$$

since

$$\underset{|u-x|>\delta}{\text{g.l.b.}}\ [x-u-(b-b_n)]^2 \geq \left(\frac{\delta}{2}\right)^{+2},$$

if $|b-b_n| \leq \delta/2$. Thus $\lim_{n\to\infty} I_2 = 0$ and $\overline{\lim_{n\to\infty}}\ |h_n(x) - \psi(x)| \leq \epsilon$. Since ϵ is arbitrary it follows that

$$\lim_{n\to\infty} h_n(x) = \psi(x).$$

13. VARIATION-DIMINISHING TRANSFORMATIONS

The formula (35), which was used to prove that $G(t)$ is non-negative, can be made to yield much more information. Suppose for the sake of definiteness that $a_1 > 0$; then for ψ in L_c^1 we set

$$f(x) = g_1 * \psi(x) = a_1 \int_x^\infty e^{a_1(x-u)}\psi(u)\, du.$$

We have at every point of continuity of ψ

$$\psi(x) = -a_1^{-1}e^{-a_1x}De^{-a_1x}f(x).$$

Now $f(x)$ is bounded and therefore $\lim_{x\to+\infty} e^{-a_1x}f(x) = 0$. Using this and Rolle's theorem, we see from the formula above that

$$\mathcal{V}[f] = \mathcal{V}[g_1 * \psi] \leq \mathcal{V}[\psi], \tag{43}$$

where $\mathcal{V}[\psi]$, the *variation* of ψ, counts the number of changes of sign of the function $\psi(x)$ in $-\infty < x < \infty$. $\mathcal{V}[\psi]$ has one of the values $0, 1, \cdots, +\infty$. Repeated application of the above argument shows that

$$\begin{aligned}\mathcal{V}[H_k * \psi] &= \mathcal{V}[g_k * g_{k-1} * \cdots * g_1 * \psi]\\ &\leq \mathcal{V}[g_{k-1} * \cdots * g_1 * \psi]\\ &\quad\vdots\\ &\leq \mathcal{V}[g_1 * \psi]\\ &\leq \mathcal{V}[\psi].\end{aligned}$$

A passage to the limit using (35) shows that

$$\mathfrak{V}[G * \psi] \leq \mathfrak{V}[\psi], \tag{44}$$

for every ψ in L_c^1.

14. SCHOENBERG'S THEOREMS

It is natural to ask what is the most general function $G(t)$ in L_p^1 such that (44) holds. In 1947 I. J. Schoenberg (see [11]) showed that a necessary and sufficient condition for (44) is that $G(t)$ be of the form

$$G(t) = \frac{d}{2\pi i} \int_{-i\infty}^{i\infty} \left[e^{bs - cs^2} \prod_k \left(1 - \frac{s}{a_k}\right) e^{s/a_k} \right]^{-1} e^{st}\, ds, \tag{45}$$

where the a_k's, b, c, and d are real and

$$\sum_k a_k^{-2} < \infty, \qquad c \geq 0. \tag{46}$$

Let us suppose that $d \geq 0$; since we can always replace $G(t)$ by $-G(t)$, this is no real restriction. With this normalization Schoenberg has proved that if $G(t)$ is of the form (45), then

$$\det\, [G(x_i - t_j)]_{i,j=1,\cdots,n} \geq 0, \tag{47}$$

whenever $-\infty < x_1 < x_2 < \cdots < x_n < \infty$, $-\infty < t_1 < t_2 < \cdots < t_n < \infty$. Moreover, the converse is true. If $G(t)$ in L_p^1 satisfies (47), then $G(t)$ is of the form (45) with $d \geq 0$.

These remarkable results show that the class of kernels studied in these sections is characterized by an important intrinsic property.

The tools that have now been developed are sufficient to carry out our program in full. In particular we obtain under extremely general assumptions an inversion formula for the convolution transform with kernel defined by (34). Moreover we obtain for each G a representation theorem analogous to Bernstein's theorem in the case of the Laplace transform (see Sec. 8), or the analogous theorem for the Stieltjes transform mentioned at the end of Sec. 9. For a detailed exposition of these results and many others we refer the reader to [8].

15. FURTHER DEVELOPMENTS

In this last section we shall indicate very briefly some further developments and also the relation of the present subject to some contiguous areas of mathematics. Let us denote by l^1 the class of functions $f(n)$ defined on the integers and such that

$$||f||_1 = \sum_{-\infty}^{\infty} |f(n)|$$

is finite. For f and g in l^1 we define

$$(f * g)(n) = \sum_{m=-\infty}^{\infty} f(n - m)g(m).$$

It is easy to see that $f * g$ again belongs to l^1 and that this operation "$*$" is commutative and associative. Let us denote by $\mathcal{V}[f]$ the number of changes of sign of $f(n)$ as n varies from $-\infty$ to ∞. In view of Schoenberg's theorem it is natural to ask what functions $G(n)$ in l^1 have the property that

$$\mathcal{V}[G * \psi] \leq \mathcal{V}[\psi] \tag{48}$$

for every ψ in l^1. This problem was first formulated by Schoenberg who conjectured that $G(n)$ must be given by a formula analogous to the formula (45) for $G(t)$ but more complicated. After partial results had been obtained by Schoenberg, Aissen, and Whitney [1], a complete solution was given by Edrei [5] in 1953. The demonstration of Edrei's result is very difficult and uses quite surprising ideas including deep results from the theory of meromorphic functions. Recently a whole class of formulas like these of Schoenberg and Edrei has been discovered by Hirschman (see for example [9]).

A square matrix

$$A = [a(i, j)], \qquad i, j = 1, \cdots, n$$

is said to be totally positive if

$$\begin{vmatrix} a(i_1, j_1) & \cdots & a(i_1, j_r) \\ a(i_2, j_1) & \cdots & a(i_2, j_r) \\ \cdot & & \\ \cdot & & \\ \cdot & & \\ a(i_r, j_1) & \cdots & a(i_r, j_r) \end{vmatrix} > 0$$

for every choice of $1 \leq i_1 < i_2 < \cdots < i_r \leq n$, $1 \leq j_1 < j_2 < \cdots < j_r < n$ and every integer r. If the above determinants are nonnegative instead of definitely positive, the matrix A is said to be totally nonnegative. See Sec. 14. Matrices with this and related properties (as well as more general operators) are studied in the book of Gantmacher and Krein [6]. Their results have spectacular applications to the theory of vibrating systems. More recently totally positive matrices have been shown to play an important role in probability theory. See for example the work of Karlin and McGregor in [10]. Many additional developments are given in a forthcoming book by Karlin.

REFERENCES

1. Aissen, M., I. J. Schoenberg, and A. Whitney, "On the generating functions of totally positive sequences," *J. Analyse Math.*, vol. 2 (1953), pp. 93–103.
2. Carslaw, H. S., and J. C. Jaeger, *Operational Methods in Applied Mathematics.* Oxford: Clarendon Press, 1941.
3. Churchill, R. V., *Modern Operational Methods in Engineering.* New York: McGraw-Hill, 1944.
4. Doetsch, G., *Theorie und Anwendung der Laplace-Transformation.* Berlin: Julius Springer, 1937.
5. Edrei, A., "On the generating function of a doubly infinite, totally positive sequence," *Trans. Amer. Math. Soc.*, vol. 74 (1953), pp. 367–383.
6. Gantmacher, F. R., and M. G. Krein, *Oszillationsmatrizen, Oszillationskerne und kleine Schwingungen mechanischer Systeme.* Berlin, Akademie-Verlag, 1960.
7. Gardner, M. F., and J. L. Barnes, *Transients in Linear Systems.* New York: John Wiley & Sons, 1942.
8. Hirschman, I. I., and D. V. Widder, *The Convolution Transform.* Princeton, N.J.: Princeton University Press, 1955.
9. Hirschman, I. I., Jr., "Variation diminishing transformations and orthogonal polynomials," *J. Analyse Math.*, vol. 9 (1961), pp. 177–193.

10. Karlin, S., and J. McGregor, "The differential equations of birth and death processes, and the Stieltjes moment problem," *Trans. Amer. Math. Soc.*, vol. 85 (1957), pp. 489–546.

11. Schoenberg, I. J., "On Pólya frequency functions. I. The totally positive functions and their Laplace transforms," *J. Analyse Math.*, vol. 1 (1951), pp. 331–374.

12. Schoenberg, I. J., "On Pólya frequency functions. II. Variation diminishing integral operators of the convolution type," *Acta Scientiarum Mathematicarum Szeged*, vol. 12 (1950), pp. 97–106.

13. Widder, D. V., *The Laplace Transform*. Princeton, N.J.: Princeton University Press, 1941.

14. Widder, D. V., *Advanced Calculus*. Englewood Cliffs, N.J.: Prentice-Hall, 1947.

A BRIEF INTRODUCTION TO THE LEBESGUE-STIELTJES INTEGRAL†

H. H. Schaefer

Among the several known approaches to the theory of integration one can distinguish, apart from the achieved generality, two basically different methods: one that defines measure on a certain class of subsets of a set S, and hence defines the integral for a corresponding class of vector valued functions on S; another that defines the integral first and hence derives the notion of measure. The former method is, historically, the earlier one; it was followed by Lebesgue in 1902. Treatments following this line are found in many textbooks covering what is usually called "the Lebesgue integral," and ranging from expositions such as [7] to comprehensive works as [8]; a general and modern approach can be found in [4]. The opposite method has entered fewer texts; the

† This article is an elaboration of an address given at the meeting of the Mathematical Association of America at the University of Oregon in June, 1959.

ones that are most easily accessible, and probably best known, are [3], [9], [10]. In [9] and [10], the Lebesgue-Stieltjes integral is introduced, starting with step-functions, in a two-stage extension modifying a procedure used by Daniell in 1918; [3] follows an approach due to F. Riesz (1920).

The present exposition, to be counted in the second group, follows in principle F. Riesz' idea. The Lebesgue-Stieltjes integral (for real-valued functions on r-dimensional euclidean space) is defined as the continuous extension of a linear form (viz., the Riemann-Stieltjes integral) on a normed vector lattice to its completion. The vector lattice C of continuous functions with compact support, a natural starting point because every function in C is Riemann-Stieltjes integrable for every additive function of intervals, has the added advantage of offering a basis with which the interested reader is likely to be familiar. The stress in the present approach is on the imbedding of the classical Lebesgue-Stieltjes theory in the framework of metric spaces, normed spaces, and vector lattices; the author has found this an opportunity to familiarize the senior undergraduate, or beginning graduate student, with the basic notions of functional analysis while treating a subject that has become an indispensable part of real analysis. The reader familiar with [2] will be aware that the present viewpoint is adopted in a very general and comprehensive manner in the cited treatise of N. Bourbaki; but [2] is not likely to provide satisfactory reading for the audience to whom this article is mainly addressed.

In order not to digress too far, the notions of metric space, normed space, vector lattice, and a few of their elementary properties and interrelations are (though briefly explained in Sec. 1) assumed as known; adequate information can be found, e.g., in [1], [5], [6]. Also, no proofs are included of the elementary properties of the Riemann-Stieltjes integral for functions in C. It should be mentioned that, in the attempt to gain the basic information on the subject as economically as possible, several important concepts (e.g., convergence in measure, absolute continuity) have been left aside. The purpose of this article is certainly accomplished if the reader desires to obtain further information;

he should then consult the references and the literature quoted there.

1. PRELIMINARIES

For the sake of reference and convenience, we shall list a few of the most elementary theorems involving metric spaces, normed vector spaces, and vector lattices, with which we shall be concerned in the sequel.

Recall that in a metric space (S, ρ) (where S is a nonempty set and ρ a distance in S), a sequence $\{x_n\}$ is a Cauchy sequence if $\lim_{m,n\to\infty} \rho(x_m, x_n) = 0$. A metric space is *complete* if every Cauchy sequence converges. Every metric space (S, ρ) can be densely imbedded into a complete metric space $(\hat{S}, \hat{\rho})$ which is uniquely determined to within isometry. The standard procedure to obtain this completion is to consider the set $\hat{S}$ of equivalence classes ξ of Cauchy sequences in S, and to define $\hat{\rho}$ on $\hat{S} \times \hat{S}$ by

$$\hat{\rho}(\xi, \eta) = \lim_{n\to\infty} \rho(x_n, y_n),$$

where $\{x_n\} \in \xi$, $\{y_n\} \in \eta$ are arbitrary representatives of the class ξ resp. η. (Here $\{x_n\}$ is *equivalent* to $\{\bar{x}_n\}$ if $\lim_{n\to\infty} \rho(x_n, \bar{x}_n) = 0$.) S then is identified with the set of those classes that contain a sequence whose range is a single element of S. A map f on (S_1, ρ_1) into (S_2, ρ_2) is *uniformly continuous* if, given any $\epsilon > 0$, $\rho_1(x, y) < \delta$ implies $\rho_2[f(x), f(y)] < \epsilon$ for some $\delta > 0$ and all $(x, y) \in S_1 \times S_1$. We shall need the following special case of the extension theorem on uniformly continuous functions:

THEOREM A: *If f is a uniformly continuous function on a metric space S_1, with values in a complete metric space S_2, then there exists a unique, uniformly continuous extension $\hat{f}$ of f to $\hat{S}_1$, with values in S_2.*

Proof: If $\hat{x} \in \hat{S}_1$ there exists a sequence $\{x_n\} \subset S_1$ such that $\hat{x} = \lim_{n\to\infty} x_n$; $\{x_n\}$ is necessarily a Cauchy sequence, and the uniform continuity of f implies that $\{f(x_n)\}$ is a Cauchy sequence in S_2 hence convergent. To define $\hat{f}$ on $\hat{S}_1$ by $\hat{f}(\hat{x}) = \lim_{n\to\infty} f(x_n)$ is unambiguous since any two sequences in S_1 with common limit $\hat{x} \in \hat{S}_1$ are equivalent so that their respective images under f are equivalent sequences in S_2, by virtue of the uniform continuity of f. To see that $\hat{f}$ is uniformly continuous, let to $\epsilon > 0$ correspond $\delta = \delta(\epsilon)$ such that $\rho_1(x, y) < \delta$ implies $\rho_2[f(x), f(y)] < \epsilon$ $((x, y) \in S_1 \times S_1)$. If $\hat{x}, \hat{y} \in \hat{S}_1$ satisfy $\rho_1(\hat{x}, \hat{y}) < \delta$ there exist sequences $\{x_n\}, \{y_n\} \subset S_1$ converging to $\hat{x}, \hat{y}$, respectively, and it follows that $\rho_1(x_m, y_n) < \delta$ for all large enough m and n; for these m and n one obtains $\rho_2[f(x_m), f(y_n)] < \epsilon$ whence we conclude that $\rho_2[\hat{f}(\hat{x}), \hat{f}(\hat{y})] \leq \epsilon$. Finally, the unicity of $\hat{f}$ is clear since any two continuous extensions of f agree on the dense subset S_1 of $\hat{S}_1$.

A (real) *normed space* (E, p) is a real vector space E and a real-valued function $x \longrightarrow p(x)$ on E such that: (1) $p(x) = 0$ implies $x = 0$. (2) $p(cx) = |c|p(x)(c \in \mathbf{R})$.† (3) $p(x + y) \leq p(x) + p(y)$. p is called a *norm*. Defining ρ on $E \times E$ by $\rho(x, y) = p(x - y)$, it is immediate that (E, ρ) is a metric space; if (E, ρ) is complete, (E, p) is called a *Banach space*. With addition and scalar multiplication suitably defined, the completion $(\hat{E}, \hat{p})$ of a normed space is a Banach space (the norm $\hat{p}$ being the continuous extension of p to $\hat{E}$).

A *linear form* on E is a real function $x \longrightarrow f(x)$ on E such that: (1) $f(x + y) = f(x) + f(y)$. (2) $f(cx) = cf(x)(c \in \mathbf{R}; x, y \in E)$. If f is a linear form on (E, p), then the following assertions are equivalent: (a) f is continuous. (b) f is uniformly continuous. (c) $|f(x)| \leq c\, p(x)$ for some $c \geq 0$ and all $x \in E$. With the aid of Theorem A, one obtains

THEOREM B: *Every continuous linear form f on a normed space E has a unique continuous extension $\hat{f}$ to $\hat{E}$; $\hat{f}$ is a linear form on $\hat{E}$.*

Let us recall the definition and some basic properties of vector lattices that will be needed below. An *ordering* (= *partial ordering*) of a set S is a binary relation "$\leq$" on S which is *reflexive* ($x \leq x$ for all $x \in S$), *transitive* ($x \leq y$ and $y \leq z$ imply $x \leq z$), and *antisymmetric* ($x \leq y$ and $y \leq x$ imply $x = y$); it is customary to write $y \geq x$ to mean $x \leq y$, and to write $x < y$ (or $y > x$) when

† $\mathbf{R}$ denotes the real number field, $\mathbf{N}$ the set of positive integers.

$x \leq y$ but $x \neq y$. A *vector lattice* is a vector space E over $\mathbf{R}$ in which an ordering is defined that satisfies the following additional axioms:

(i) *$x \leq y$ implies $x + z \leq y + z$ for all $z \in E$ (translation invariance).*

(ii) *$x \leq y$ implies $\lambda x \leq \lambda y$ for all real numbers $\lambda > 0$.*

(iii) *For each pair $(x, y) \in E \times E$,* sup (x, y) *and* inf (x, y) *exist in E.*

Here sup (x, y) is an element $u \in E$ such that $u \geq x, u \geq y$, and such that the relations $x \leq z, y \leq z$ imply $u \leq z$ for all $z \in E$; inf (x, y) is defined in an analogous fashion, and one has inf $(x, y) = -\sup(-x, -y)$ by virtue of (i). sup (x, y) and inf (x, y) are unique and related by the identity sup $(x, y) +$ inf $(x, y) = x + y$; in particular, for any element x of the vector lattice E one defines $x^+ = \sup(x, 0)$, $x^- = \sup(-x, 0)$ so that $x = x^+ - x^-$. The *absolute* $|x|$ of $x \in E$ is defined to be sup $(x, -x)$; one obtains $|x| = x^+ + x^-$, $|x + y| \leq |x| + |y|$, and $|\lambda x| = |\lambda|\,|x|$ for all $x \in E$, $y \in E$, $\lambda \in \mathbf{R}$. Proofs of these results can be found, for instance, in [1]; in the case of vector lattices of real functions (or classes of real functions) with which we shall exclusively be concerned, they are almost self-evident.

A *vector sublattice* of a vector lattice E is a vector subspace closed under the lattice operations; in other words, a linear subspace F of E such that $x \in F$, $y \in F$ implies sup $(x, y) \in F$ (where the sup is to be taken in E) or, equivalently, such that $x \in F$ implies $|x| \in F$. A vector sublattice F of E is *solid* if $x \in F$, $y \in E$ and $|y| \leq |x|$ imply $y \in F$.

Examples

1. Let X denote the space $\mathbf{R}^r$ (euclidean r-space), C the vector space of real-valued continuous functions f vanishing outside some compact set depending on f. Under the ordering defined by "$f \leq g$ if $f(t) \leq g(t)$ for all $t \in X$," C is a vector lattice; for f, $g \in C$ sup (f, g) is the function which is, at each $t \in X$, the numerical upper bound sup $(f(t), g(t))$.

2. The vector space $\mathbf{R}^X$ of all real-valued functions f on X (X

an arbitrary set $\neq \emptyset$) is a vector lattice under "$f_1 \leq f_2$ if $f_1(t) \leq f_2(t)$ for all $t \in X$." The space $B(X)$ of bounded real functions on X is a solid vector sublattice of $\mathbf{R}^X$. By contrast, if $X = \mathbf{R}^r$ (Example 1) then C is a vector sublattice of $B(X)$ and of $\mathbf{R}^X$ but not a solid sublattice of either one.

3. Let $BV(\mathbf{R})$ be the vector space of all real functions h on $\mathbf{R}$ that are of bounded variation on every finite interval and such that $h(0) = 0$. Under the ordering "$h_1 \leq h_2$ if $h_2 - h_1$ is nondecreasing" $BV(\mathbf{R})$ is a vector lattice. (To verify axiom (iii), observe that for $t \geq 0$, h^+ is given by

$$h^+(t) = \sup \sum_{k=1}^{n} [h(t_k) - h(t_{k-1})],$$

where $\{[t_{k-1}, t_k]: k = 1, \cdots, n\}$ is any finite family of nonoverlapping subintervals of the interval $[0, t]$; an analogous representation is valid for $t < 0$.)

If E is a vector lattice and H a solid vector sublattice of E, the quotient space E/H can be made into a vector lattice in a natural way; if ϕ is the canonical map (which orders to each $x \in E$ its equivalence class $[x]$ in E/H), the definition "$[x] \leq [y]$ if there exist $x \in [x]$, $y \in [y]$ such that $x \leq y$" defines a lattice structure on E/H for which ϕ is a lattice homomorphism (that is, $\phi(\sup(x, y)) = \sup(\phi(x), \phi(y))$). The relation $[x] \leq [y]$ is clearly reflexive; to show that it is transitive, let $[x] \leq [y]$ and $[y] \leq [z]$. There exist elements $x_1 \in [x]$, $y_1, y_2 \in [y]$, $z_2 \in [z]$ such that $x_1 \leq y_1$, $y_2 \leq z_2$ and we have

$$x_1 \leq y_1 \leq y_2 + |y_1 - y_2| \leq z_2 + |y_1 - y_2|,$$

which shows that $[x] \leq [z]$ since $|y_1 - y_2| \in H$. In a simiiar way it follows that the relation $\leq$ on E/H is antisymmetric (one needs here that H is a solid sublattice). It remains to be seen that ϕ is a lattice homomorphism, that is, $[\sup(x, y)] = \sup([x], [y])$ for all $x, y \in E$. In fact, if $z = \sup(x, y)$ it is clear that $[z] \geq [x]$ and $[z] \geq [y]$; suppose that $[w] \geq [x]$ and $[w] \geq [y]$ for some $[w] \in E/H$. There exist $w_1, w_2 \in [w]$ such that $w_1 \geq x$, $w_2 \geq y$; let $w_3 = \sup(w_1, w_2)$. It follows that $w_1 \leq w_3 \leq w_1 + |w_2 - w_1|$, since the last term is $\geq w_2$. Therefore

$$0 \leq w_3 - w_1 \leq |w_2 - w_1| \in H;$$

thus since H is solid we have $w_3 \in [w]$, and since clearly $w_3 \geq z$ it follows that $[w] \geq [z]$, which proves our assertion.

If $(E, p, \leq)$ is a normed space and a vector lattice, we shall say that E is a *normed lattice* if $|y| \leq |x|$ implies $p(y) \leq p(x)$ (hence, in particular, $p(x) = p(|x|)$ for all $x \in E$). It follows from this that the lattice operations are continuous (even uniformly continuous); for instance, from $x = y + (x - y)$ one derives $x^+ \leq y^+ + |x - y|$ and, interchanging x and y, $|x^+ - y^+| \leq |x - y|$, which implies that $p(x^+ - y^+) \leq p(x - y)$ and hence that $x \longrightarrow x^+$ is uniformly continuous on E into E. Clearly the same holds for $x \longrightarrow x^-$ and $x \longrightarrow |x|$; more generally, by axiom (i) one has

$$\sup(x, y) = x + \sup(0, y - x) = x + (y - x)^+,$$

and it follows readily from this and the preceding that

$$|\sup(x, y) - \sup(x_1, y_1)| \leq |x - x_1| + |y - y_1|.$$

This implies the uniform continuity of $(x, y) \longrightarrow \sup(x, y)$ on $E \times E$ into E. As a particular consequence of the continuity of the lattice operations, we note that the positive cone $\{x : x \geq 0\}$ of E is closed; in fact, this cone is the inverse image of the closed set $\{0\} \subset E$ under $x \longrightarrow x^-$.

A *Banach lattice* is a normed lattice which is complete as a metric space (hence, in particular, is a Banach space). Using the uniform continuity of the lattice operations in a normed lattice $(E, p, \leq)$, the completion $(\hat{E}, \hat{p})$ can be made into a Banach lattice through continuous extension of $(x, y) \longrightarrow \sup(x, y)$ to $\hat{E} \times \hat{E}$; for the following theorem, we prefer a somewhat different proof since our proof will supply a pattern to be followed in Sec. 3 below and, in addition, provide a general example for the construction of quotient lattices described above. A linear form f on a normed lattice is called *positive* (respectively, *strictly positive*) if $x \geq 0$ implies $f(x) \geq 0$ (respectively, if $x > 0$ implies $f(x) > 0$).

THEOREM C: *With respect to the continuous extensions of the lattice operations and the norm, the completion of a normed lattice E is a Banach lattice $\hat{E}$. If f is a continuous positive linear form on E, its continuous extension $\hat{f}$ to $\hat{E}$ is a positive linear form on $\hat{E}$.*

Proof: The set E_1 of all Cauchy sequences $\xi = \{x_n\}_{n \in \mathbf{N}}$ is clearly a vector space under coordinatewise addition and scalar multiplication; it is simple to verify that E_1 is a vector lattice under "$\xi \leq \eta$ if $x_n \leq y_n$ for all $n \in \mathbf{N}$." Let N be the vector subspace of E_1 formed by all sequences converging to 0 in E; we define, in the standard manner, $\hat{E}$ to be the quotient E_1/N, and the norm $\hat{p}$ of $\hat{E}$ is defined by $p([\xi]) = \lim_{n\to\infty} p(x_n)$ where $\{x_n\}$ is any element of $[\xi] \in E_1/N$. Under the ordering of E_1 defined above, N is clearly a solid sublattice of E_1, and hence the canonical map $E_1 \longrightarrow E_1/N$ defines a vector lattice structure on $\hat{E}$. Whenever $|[\xi]| \leq |[\eta]|$, one has

$$\begin{aligned} \hat{p}([\xi]) &= \lim_{n\to\infty} p(x_n) = \lim_{n\to\infty} p(|x_n|) \\ &\leq \lim_{n\to\infty} p(|y_n|) = \lim_{n\to\infty} p(y_n) = \hat{p}([\eta]) \end{aligned}$$

for $\{x_n\} \in [\xi]$, $\{y_n\} \in [\eta]$; thus $(\hat{E}, \hat{p}, \leq)$ is a Banach lattice. Now consider the mapping ψ that orders to each $x \in E$ the unique equivalence class $[\xi]$ that contains the sequence $\{x_n\}$ for which $x_n = x (n \in \mathbf{N})$; it is immediate that ψ is an isomorphism of $(E, p, \leq)$ into $(\hat{E}, \hat{p}, \leq)$. Moreover, it follows from a standard argument that $\psi(E)$ is dense in $\hat{E}$; this implies, in particular, that $\hat{p}$ is the continuous extension of p, and hence that the lattice operations in $\hat{E}$ are the continuous extensions of those in E (when E is identified with $\psi(E)$). Verification of the final assertion can be left to the reader.

We conclude this preliminary section with a covering lemma for open sets in euclidean r-space.

THEOREM D: *If G is an open subset of $\mathbf{R}^r$, there exists a countable family $\{I_n : n \in \mathbf{N}\}$ of closed intervals with mutually disjoint interior, such that $G = \bigcup_1^\infty I_n$ and each $x \in G$ is interior to the union of a suitable finite subfamily.*

Proof: Denote by Q_k $(k = 0, 1, \cdots)$ the subset of $\mathbf{R}^r$ such that $a = (a_1, \cdots, a_r) \in Q_k$ if and only if the numbers $2^k a_\rho$ $(\rho = 1, \cdots, r)$ are integral; clearly $\bigcup_0^\infty Q_k$ is dense in $\mathbf{R}^r$. If G is a given open subset of $\mathbf{R}^r$, denote by $\mathfrak{W}_0$ the set of all intervals $I = \{x: a_\rho \leq x_\rho \leq a_\rho + 1, \rho = 1, \cdots, r\}$ such that $a \in Q_0$ and $I \subset G$. The countable (possibly empty) sets $\mathfrak{W}_k$ $(k \geq 1)$ of closed intervals are now defined by induction: Assuming that $\mathfrak{W}_l$ $(l = 0, \cdots, k - 1)$ have been determined, let $\mathfrak{W}_k$ be the set of all intervals $I = \{x: a_\rho \leq x_\rho \leq a_\rho + 2^{-k}, \rho = 1, \cdots, r\}$ such that $a \in Q_k$, $I \subset G$ and I is not contained in any interval belonging to $\mathfrak{W}_l$ $(l = 0, \cdots, k-1)$. $\mathfrak{W} = \bigcup_0^\infty \mathfrak{W}_k$ is a countable set of intervals with mutually disjoint interior and such that $\bigcup \{I: I \in \mathfrak{W}\} \subset G$. Let $x \in G$ be fixed; since G is open, there exists $\epsilon > 0$ such that $d(x, z) \geq 2\epsilon$ whenever $z \notin G$ (d the euclidean distance). Let k_0 be the smallest integer ≥ 0 such that $\sqrt{r}\, 2^{-k_0} < \epsilon$, and let $G_1 = \{y \in G: d(x, y) < \epsilon\}$. It follows that each $y \in G_1$ is contained in an interval $I \in \mathfrak{W}_k$ where $k \leq k_0$, and the set of all intervals intersecting G_1 and belonging to some $\mathfrak{W}_k$ where $k \leq k_0$ is finite, for no two of them have a common interior point while their union J is contained in the set $\{z \in \mathbf{R}^r: d(x, z) < (1 + \epsilon)\sqrt{r}\}$. On the other hand, $J \supset G_1$; hence x is interior to J, which completes the proof.

2. THE RIEMANN-STIELTJES INTEGRAL

In all following sections we shall denote by X the r-dimensional euclidean space $\mathbf{R}^r$, where r is an arbitrary fixed integer ≥ 1. By an *interval* in X we mean a subset $I = \{x: a_\rho \leq x_\rho \leq b_\rho, \rho = 1, \cdots, r\}$; hence "interval" means what is usually called a closed interval. An *open interval* is a set of the form $\{x: a_\rho < x_\rho < b_\rho; \rho = 1, \cdots, r\}$. If $\mathfrak{J}$ is the family of all intervals in X, a non-negative real function α on $\mathfrak{J}$ is called (*finitely*) *additive* if $\alpha(I) = \Sigma_\nu \alpha(I_\nu)$ whenever $I = \bigcup I_\nu$ is the union of finitely many sub-intervals I_ν with mutually disjoint interior. We note that $\alpha(I) = 0$ if I has empty interior since, in this case, we must have $\alpha(I) = \alpha(I) + \alpha(I)$. If $r = 1$, every nondecreasing real function $x \longrightarrow f(x)$ defines an additive function α of intervals by virtue of $\alpha(I) = f(b) - f(a)$ whenever $I = [a, b]$ $(b \geq a)$. If $r > 1$, a similar construction can be carried out. For example, let $r = 2$

and suppose f is a real-valued function on $\mathbf{R}^2$ such that, whenever $b_1 \geq a_1$ and $b_2 \geq a_2$, it is true that $f(b_1, b_2) - f(a_1, b_2) \geq f(b_1, a_2) - f(a_1, a_2)$; then the difference of the left- and right-hand terms defines $\alpha(I)$ for $I = \{x: a_\rho \leq x_\rho \leq b_\rho; \rho = 1, 2\}$. (Every twice-differentiable f such that $\partial^2 f/\partial x_1 \, \partial x_2 \geq 0$ furnishes an example; $f(x_1, x_2) = x_1 x_2$ yields the elementary area of I.)

We shall denote by C the vector lattice of continuous real functions on X with compact support (Sec. 1, Example 1). Let $f \in C$, and let I be an interval such that $f(x) \neq 0$ implies $x \in I$. Let Σ denote a set of pairs (I_κ, x^κ) $(\kappa = 1, \cdots, k)$ such that $x^\kappa \in I_\kappa$ for all κ and $I = \bigcup_{\kappa=1}^{k} I_\kappa$ while no two I_κ have a common interior point. If $\delta(I_\kappa)$ is the length of the diagonal of I_κ, $\delta = \max_{1 \leq \kappa \leq k} \delta(I_\kappa)$ is called the *mesh* of Σ. The number

$$S(f; \Sigma) = \sum_{\kappa=1}^{k} f(x^\kappa)\alpha(I_\kappa)$$

is known as the *Riemann-Stieltjes sum* of f (with respect to Σ). Keeping I fixed and considering a sequence (Σ_n) such that mesh $\Sigma_n \longrightarrow 0$, it is easy to establish that $S(f; \Sigma_n)$ converges; we write

$$\lim_{n \to \infty} S(f; \Sigma_n) = \int f \, d\alpha,$$

the limit being independent of the particular sequence (Σ_n) and the interval I chosen, as long as mesh $\Sigma_n \longrightarrow 0$ and $f(x) = 0$ on the boundary and outside of I. $\int f \, d\alpha$ is called the *Riemann-Stieltjes integral* of f (with respect to α). If $\alpha(I)$ is taken to be the elementary volume of I, $\int f \, d\alpha$ is called the *Riemann integral* of $f \in C$. We remark that for the limit $\lim S(f, \Sigma_n)$ to exist, it is not necessary to demand that $f(x) = 0$ on the boundary and exterior of I. Let $I = \bigcup_{\lambda=1}^{l} I_\lambda$ be a subdivision of I into intervals with mutually disjoint interior, and let $S(f; \Sigma_{n,\lambda})$ be any Riemann-Stieltjes sum of f over I_λ whose intervals are the nonempty intersections with I_λ of the intervals of Σ_n. If f is continuous on I, and if mesh $\Sigma_n \longrightarrow 0$, the limits

$$\lim_{n \to \infty} S(f; \Sigma_{n,\lambda}) = \int_{I_\lambda} f \, d\alpha, \qquad \lambda = 1, \cdots, l,$$

exist, and obviously one has

$$\int_I f\,d\alpha = \sum_{\lambda=1}^{l} \int_{I_\lambda} f\,d\alpha.$$

Hence:

If α is an arbitrary, finitely additive function of intervals, and $f \geq 0$ is a fixed continuous real function on X, then the function

$$I \longrightarrow \int_I f\,d\alpha$$

is a finitely additive function of intervals in X.

We note the following properties of $\int f\,d\alpha$:

1. $\int (f+g)\,d\alpha = \int f\,d\alpha + \int g\,d\alpha, f, g \in C$.
2. $\int cf\,d\alpha = c \int f\,d\alpha, f \in C, c \in \mathbf{R}$.
3. $|\int f\,d\alpha| \leq \int |f|\,d\alpha, f \in C$.

These properties are equivalent to:

The mapping $f \longrightarrow \int f\,d\alpha$ is a positive linear form on the vector lattice C.

Letting $\dot{p}_\alpha(f) = \int |f|\,d\alpha$, it follows from properties 1 and 3 that $\dot{p}_\alpha(f+g) \leq \dot{p}_\alpha(f) + \dot{p}_\alpha(g)$, and from property 2 that $\dot{p}_\alpha(cf) = |c|\dot{p}_\alpha(f)$; hence $\dot{p}_\alpha$ has all properties of a norm except that $\dot{p}_\alpha(f) = 0$ might not imply $f = 0$. However, the set $\{f: \dot{p}_\alpha(f) = 0\}$ is a solid vector sublattice N_α of C. Denoting by C_α the quotient space C/N_α of equivalence classes $[f]$ of C mod N_α, the map $p_\alpha: [f] \longrightarrow \dot{p}_\alpha(f)$ is a norm on C_α. It is easy to verify (see Sec. 1) that C_α is a lattice under the ordering defined by $[f] \leq [g]$ if $f \leq g$ for some $f \in [f], g \in [g]$, and that $[f] \longrightarrow \varphi([f]) = \int f\,d\alpha$ is a linear form on C_α.

(C_α, p_α) is a normed vector lattice, and *φ is a positive, continuous linear form on C_α.* For $|\varphi([f])| = |\int f\,d\alpha| \leq \int |f|\,d\alpha = p_\alpha([f])$. By Theorem C, the completion of (C_α, p_α), i.e., the set of equivalence classes of Cauchy sequences modulo null sequences, is a Banach lattice. We shall denote this space by $\hat{C}_\alpha$, and the continuous extension of the linear form φ to $\hat{C}_\alpha$, by $\hat{\varphi}$.

3. THE SPACES $L(\alpha)$

It is the purpose of this section to identify the elements of $\hat{C}_\alpha$ as certain classes of finite, real-valued functions on X. For convenience, we shall say that a sequence $\{f_n\}$ in C is a p_α-Cauchy sequence if $\lim_{m,n\to\infty} \int |f_m - f_n|\, d\alpha = 0$, i.e., if the sequence $\{[f_n]\}$ is a Cauchy sequence in (C_α, p_α). Throughout this section, α is a fixed (non-negative), finitely additive function of intervals in X.

DEFINITION 1: *$A \subset X$ is ϵ-small if there exists a countable family $\{I_n : n \in \mathbf{N}\}$ of intervals with $\sum_1^\infty \alpha(I_n) \leq \epsilon$, and such that every $x \in A$ is interior to the union of some finite subfamily of $\{I_n\}$. A set which is ϵ-small for every $\epsilon > 0$ is called a zero set.*†

It follows from this definition that the union A of a countable family $\{A_n : n \in \mathbf{N}\}$ of zero sets is a zero set. In fact, given $\epsilon > 0$ there exist intervals $I_{n,m}$ such that $\sum_{m=1}^{\infty} \alpha(I_{n,m}) \leq \epsilon/2^n$ and such that each $x \in A_n$ is interior to a finite union $\bigcup_{k=1}^{p} I_{n,k}$; the family $\{I_{n,m} : (n, m) \in \mathbf{N} \times \mathbf{N}\}$ is then a covering of A with α-sum $\leq \epsilon$. Since this can be arranged for each $\epsilon > 0$, A is a zero set.

If a statement P on points x is true for all $x \in X$ with the exception of a zero set, we say that P is true *almost everywhere* (*a.e.*). Denote by $\mathbf{R}^X$ the vector space of all (finite) real-valued functions on X.

DEFINITION 2: *A sequence $\{f_n\}$ in C converges to an $f \in \mathbf{R}^X$ almost uniformly (a.u.) if to each $\epsilon > 0$, there exists an ϵ-small set X_ϵ such that $\lim_{n\to\infty} f_n(x) = f(x)$ uniformly in $X - X_\epsilon$.*

We note that if $f_n \longrightarrow f$ a.u., then $f_n \longrightarrow f$ a.e., the converse being, in general, false.

LEMMA 1: *If $f \in C$ and $\int |f|\, d\alpha \leq \epsilon^2 (\epsilon > 0)$, then $|f(x)| \leq \epsilon$ for $x \in X - X_\epsilon$ where X_ϵ is ϵ-small.*

---

† If reference to α is desired, we shall use the terms "ϵ-small $[\alpha]$" and "zero set $[\alpha]$."

Proof: Let $X_\epsilon = \{x: |f(x)| > \epsilon\}$. X_ϵ is an open (bounded) subset of X, hence by Theorem D (Sec. 1) the union of countably many intervals $\{I_n: n \in \mathbf{N}\}$ of the required type. If $\sum_{n=1}^{\infty} \alpha(I_n) > \epsilon$, then there is a finite subset $\{I_n: n = 1, \cdots, m\}$ such that $\sum_{n=1}^{m} \alpha(I_n) > \epsilon$. Since obviously, then,

$$\int |f| \, d\alpha \geq \epsilon \sum_{n=1}^{m} \alpha(I_n) > \epsilon^2,$$

we obtain a contradiction. Hence $\sum_{n=1}^{\infty} \alpha(I_n) \leq \epsilon$, i.e., X_ϵ is ϵ-small.

PROPOSITION 1: *Every p_α-Cauchy sequence in C contains a subsequence that converges a.u. to some $f \in \mathbf{R}^X$.*

Proof: Let $\{f_n\}$ be a subsequence of the given sequence such that

$$\int |f_{n+1} - f_n| \, d\alpha \leq \left(\frac{1}{2^{n+1}}\right)^2.$$

Let $X_n \subset X$ be the set in which $|f_{n+1}(x) - f_n(x)| \leq 2^{-(n+1)}$ fails to be true; by Lemma 1, X_n is $2^{-(n+1)}$-small. Letting $Y_n = \bigcup_{\nu=n}^{\infty} X_\nu$, it follows that Y_n is 2^{-n}-small; on the other hand, $\{f_n\}$ converges uniformly outside of every Y_n. If $Y = \bigcap_1^{\infty} Y_n$, Y is a zero set; $f_n(x)$ converges in $X - Y$. Defining $x \longrightarrow f(x)$ by $f(x) = \lim_{n\to\infty} f_n(x)$ if $x \in X - Y$, and by $f(x) = 0$ if $x \in Y$, we have shown that $f_n \longrightarrow f$ a.u.

PROPOSITION 2: *If* $\{f_n\}$ (respectively $\{g_n\}$) *is a sequence in C converging a.e. to f* (resp. g), *and if* $\int |f_n - g_n| \, d\alpha \longrightarrow 0$, *then* $f(x) = g(x)$ *a.e.*

Proof: By hypothesis, $\int |f_n - g_n| \, d\alpha = \epsilon_n \longrightarrow 0$. If

$$Z_n = \{x: |f_n(x) - g_n(x)| > \sqrt{\epsilon_n}\},$$

then Z_n is $\sqrt{\epsilon_n}$-small by Lemma 1. It follows that $Z = \bigcup_{k=1}^{\infty} \bigcap_{n=k}^{\infty} Z_n$

is a zero set. Let X_1 (resp. X_2) be the zero set in which $f_n(x)$ (resp. $g_n(x)$) fails to converge; $Y = X_1 \cup X_2 \cup Z$ is a zero set. On the other hand, since

$$|f(x) - g(x)| \leq |f(x) - f_n(x)| + |f_n(x) - g_n(x)| + |g_n(x) - g(x)|$$

holds for all $n \in \mathbf{N}$ and $x \in X$, we have $\{x : f(x) \neq g(x)\} \subset Y$ and the proof is complete.

PROPOSITION 3: *Let $\{f_n\}$ be a p_α-Cauchy sequence in C converging to 0 a.u. Then $\int |f_n| \, d\alpha \longrightarrow 0$, i.e., $\{f_n\}$ is a p_α-null sequence.*

Proof: Without loss of generality, assume that $f_n \neq 0$ for all $n \in \mathbf{N}$ and that $\alpha(I) \neq 0$ for some interval $I \subset X$. Let $\epsilon > 0$ be given; there exists an n_0 such that $\int |f_n - f_{n_0}| \, d\alpha < \epsilon/6$ if $n > n_0$. Set $M = \sup \{|f_{n_0}(x)| : x \in X\}$ and let I_0 be an interval such that $f_{n_0}(x) \neq 0$ implies $x \in I_0$, and such that $\alpha(I_0) > 0$. Further let I_m be an interval containing I_0, and such that $f_m(x) \neq 0$ implies $x \in I_m$. Since $f_m \longrightarrow 0$ a.u., there exists an $(\epsilon/6M)$-small set X_ϵ, and an $m_0 \geq n_0$, such that

$$|f_m(x)| \leq \frac{\epsilon}{6\alpha(I_0)}, \qquad \text{if } m \geq m_0, x \in X - X_\epsilon.$$

Let now $m \geq m_0$ be fixed. Since $I \longrightarrow \int_I |f_m| \, d\alpha$ is additive on intervals (Sec. 2), we may write

$$\int |f_m| \, d\alpha = \int_{I_0} |f_m| \, d\alpha + \int_{I_m - I_0} |f_m| \, d\alpha, \tag{1}$$

where $I_m - I_0$ stands for a union of subintervals of I_m. Since $f_{n_0}(x) = 0$ for $x \in I_m - I_0$, the second term is equal to $\int_{I_m - I_0} |f_m - f_{n_0}| \, d\alpha$, and hence $< \epsilon/6$. To estimate the first term, we choose a subdivision of I_0 into intervals I_λ $(\lambda = 1, \cdots, l)$ with mutually disjoint interior, and mesh so small that

$$\left|\sum_1^l |f_m(x^\lambda)| \alpha(I_\lambda) - \int_{I_0} |f_m| \, d\alpha\right| < \frac{\epsilon}{6} \tag{2}$$

and

$$\left|\sum_1 |f_m(x^\lambda) - f_{n_0}(x^\lambda)| \alpha(I_\lambda) - \int_{I_0} |f_m - f_{n_0}| \, d\alpha\right| < \frac{\epsilon}{6} \tag{3}$$

for any choice of $x^\lambda \in I_\lambda (\lambda = 1, \cdots, l)$. (Such a subdivision exists owing to the uniform continuity of f_{n_0} and f_m on I_0.) We select among the collection $\{I_\lambda\}$ one group, $\{I'_\lambda\}$, with the property that $x \in I'_\lambda$ implies $|f_m(x)| > \epsilon/6\alpha(I_0)$, and put the remaining intervals into a group $\{I''_\lambda\}$. For those λ, $1 \leq \lambda \leq l$, for which I_λ is in the second group, we pick an $x^\lambda \in I''_\lambda$ such that $|f_m(x^\lambda)| \leq \epsilon/6\alpha(I_0)$; for the remainder, let x^λ be any point in I'_λ. We obtain

$$\Sigma\, |f_m(x^\lambda)|\alpha(I''_\lambda) \leq \frac{\epsilon}{6\alpha(I_0)} \Sigma\, \alpha(I''_\lambda) \leq \frac{\epsilon}{6},$$

and

$$(4) \quad \Sigma\, |f_m(x^\lambda)|\alpha(I'_\lambda) \leq \Sigma |f_{n_0}(x^\lambda)|\alpha(I'_\lambda) + \Sigma\, |f_m(x^\lambda) - f_{n_0}(x^\lambda)|\alpha(I'_\lambda).$$

Obviously, we have $\cup\, I'_\lambda \subset X_\epsilon$. Hence, by Definition 1, there exists a countable collection $\{J_n\}$ of intervals such that $\sum_1^\infty \alpha(J_n) \leq \epsilon/6M$ and each $x \in X_\epsilon$ is interior to some finite union of elements in $\{J_n\}$. Thus, $\cup\, I'_\lambda$ being compact, $\cup\, I'_\lambda$ is contained in the union of finitely many of the J_n, and from the finite additivity of α it follows that $\Sigma\, \alpha(I'_\lambda) \leq \epsilon/6M$. Since M is the upper bound of $|f_{n_0}|$, the first right-hand term in (4) is $\leq \epsilon/6$. The second term is not larger than $\sum_{\lambda=1}^{l} |f_m(x^\lambda) - f_{n_0}(x^\lambda)|\alpha(I_\lambda)$ which, by (3), is in turn not larger than

$$\int_{I_0} |f_m - f_{n_0}|\, d\alpha + \frac{\epsilon}{6} \leq \int |f_m - f_{n_0}|\, d\alpha + \frac{\epsilon}{6} < \frac{1}{3}\,\epsilon.$$

(2) implies now $\int_{I_0} |f_m|\, d\alpha < \frac{5}{6}\epsilon$. Hence, by (1), we have $\int |f_m|\, d\alpha < \epsilon$ for an arbitrary $m \geq m_0$ and the proof is finished.

Let us denote by $\mathfrak{L}(\alpha)$ the subset of $\mathbf{R}^X$ of all a.u. limits of p_α-Cauchy sequences in C. It is obvious that if $f_n \longrightarrow f$, $g_n \longrightarrow g$ a.u. with $f_n, g_n \in C (n \in N)$, then: (a) $(f_n + g_n) \longrightarrow (f + g)$ a.u., (b) $cf_n \longrightarrow cf$ a.u. $(c \in \mathbf{R})$, and (c) $\sup\,(f_n, g_n) \longrightarrow \sup\,(f, g)$ a.u. (resp. $\inf\,(f_n, g_n) \longrightarrow \inf\,(f, g)$ a.u.). Statement (c) follows from the inequality

$$|\sup\,(a, b) - \sup\,(a_1, b_1)| \leq |a - a_1| + |b - b_1|,$$

valid for any four real numbers a, a_1, b, b_1 (Sec. 1). Hence $\mathfrak{L}(\alpha)$ is a vector sublattice of $\mathbf{R}^X$. We have to go one step further. If M_α denotes (the subset of $\mathfrak{L}(\alpha)$ of) all $f \in \mathbf{R}^X$ such that $f(x) = 0$ a.e., then M_α is a vector sublattice of $\mathfrak{L}(\alpha)$ which is solid in $\mathbf{R}^X$, i.e., such that $f \in M_\alpha$, $g \in \mathbf{R}^X$ and $0 \leq g \leq f$ imply $g \in M_\alpha$. From this it follows (Sec. 1) that the quotient $\mathfrak{L}(\alpha)/M_\alpha$ is itself a vector lattice, with vector space and lattice operations defined in the natural way. We shall denote this quotient lattice by $L(\alpha)$.

Now we define a mapping χ of the completion $\hat{C}_\alpha$ of (C_α, p_α) onto $L(\alpha)$: Let to each equivalence class $\xi \in \hat{C}_\alpha$ (modulo null sequences) of p_α-Cauchy sequences of functions in C, correspond the set of all $f \in \mathbf{R}^X$ that are a.u. limits of some sequence in ξ.

It follows from Proposition 1 that for every $\xi \in \hat{C}_\alpha$, $\chi(\xi)$ is nonempty. By Proposition 2, $\chi(\xi)$ is contained in some element of $L(\alpha)$, and by the definition of a.u. convergence, actually $\chi(\xi) \in L(\alpha)$. Then, obviously, $\chi(\hat{C}_\alpha) = L(\alpha)$. By the above remarks (a) through (c) χ is linear and preserves lattice operations. Finally, by Proposition 3, χ is one-to-one. Hence we have proved:

PROPOSITION 4: *The mapping χ is a vector lattice isomorphism of $\hat{C}_\alpha$ onto $L(\alpha)$.*

4. THE LEBESGUE-STIELTJES INTEGRAL. CONVERGENCE THEOREMS

By virtue of Proposition 4 we may, and shall, identify the space $L(\alpha)$ as the completion of (C_α, p_α). Let $-\hat{\varphi}$ denote the unique continuous extension (Theorem B) of the linear form $[f] \longrightarrow \int f\,d\alpha$ on (C_α, p_α) to $L(\alpha)$. Then for each $f \in \mathfrak{L}(\alpha)$, contained in the class $[f] \in L(\alpha)$, we define the real number $\hat{\varphi}([f])$ to be the *Lebesgue-Stieltjes integral* $\int f\,d\alpha$ of f with respect to α. The elements of $\mathfrak{L}(\alpha)$ are called α-summable (or briefly summable if no confusion is likely to occur). If α is the function of intervals in X for which $\alpha(I) =$ elementary volume of I, $\int f\,d\alpha$ is called the *Lebesgue integral* of $f \in \mathfrak{L}(\alpha)$.

It follows from this definition that if f is α-summable and $f(x) = g(x)$ a.e. in X, then so is g and $\int f\,d\alpha = \int g\,d\alpha$. Further, if f is continuous and of compact support (i.e., $f \in C$), then f is

α-summable for every α and $\int f\,d\alpha$ coincides with the Riemann-Stieltjes integral of f.†

THEOREM 1: *Under the norm* $[f] \longrightarrow \int |f|\,d\alpha$, $L(\alpha)$ *is a Banach lattice, of which* $\chi(C_\alpha)$ *is a dense sublattice, and* $[f] \longrightarrow \int f\,d\alpha$ *is a strictly positive linear form on* $L(\alpha)$.

Proof: Since $[f] \longrightarrow \int |f|\,d\alpha = \hat{\varphi}([|f|])$ is the continuous extension $\hat{p}_\alpha$ to $L(\alpha)$ of the norm p_α on C_α, the first assertion is an immediate consequence of Theorem C. It remains to prove that $[f] \geq 0$ and $\int f\,d\alpha = 0$ imply $[f] = 0$. But since on the positive cone $\{[f]: [f] \geq 0\}$ of $L(\alpha)$, $\hat{\varphi}$ coincides with the norm $\hat{p}_\alpha$, it is also clear that $\hat{\varphi}$ is strictly positive.

We remark that Theorem 1 contains the following statements: If f, g are summable, $c \in \mathbf{R}$, then so are $f + g$, cf, $\sup(f, g)$ and $\inf(f,g)$, and we have

1. $\int (f + g)\,d\alpha = \int f\,d\alpha + \int g\,d\alpha$.
2. $\int cf\,d\alpha = c \int f\,d\alpha$.
3. $|\int f\,d\alpha| \leq \int |f|\,d\alpha$.

These properties are all shared by the Riemann-Stieltjes integral defined in Sec. 2. The situation is different with this property which expresses the completeness of $L(\alpha)$, and is the key to the powerful convergence theorems that will be proved shortly: If $\{f_n\}$ is a sequence in $\mathfrak{L}(\alpha)$ such that $\int |f_m - f_n|\,d\alpha \longrightarrow 0$ $(m, n \longrightarrow \infty)$, there exists an $f \in \mathfrak{L}(\alpha)$ such that $\int |f_n - f|\,d\alpha \longrightarrow 0$ (and hence $\int f_n\,d\alpha \longrightarrow \int f\,d\alpha$). Further, for each $f \in \mathfrak{L}(\alpha)$ and $\epsilon > 0$ there exists an $h \in C$ such that $\int |f - h|\,d\alpha < \epsilon$.

It is now easy to prove that Propositions 1–3 remain valid for sequences in $\mathfrak{L}(\alpha)$ that are $\hat{p}_\alpha$-Cauchy sequences, i.e., for which $\int |f_n - f_m|\,d\alpha \longrightarrow 0$. To indicate the method of proof, we establish a slightly different form of Lemma 1 for elements in $\mathfrak{L}(\alpha)$.

---

† We remark that C is the largest subspace of $\mathbf{R}^X$ such that all elements are Riemann-Stieltjes integrable for every α. (A bounded $f \in \mathbf{R}^X$ with compact support is Riemann-Stieltjes integrable $[\alpha]$ if $\lim S(f; \Sigma_n)$, of Sec. 2, exists for every sequence (Σ_n) with mesh $\Sigma_n \longrightarrow 0$.)

LEMMA 1': *If* $f \in \mathfrak{L}(\alpha)$ *and* $\int |f|\, d\alpha < \epsilon^2$ $(\epsilon > 0)$, *then* $|f(x)| < \epsilon$ *in* $X - X_\epsilon$ *where* X_ϵ *is* ϵ-*small.*

Proof: Let $\int |f|\, d\alpha = \delta$ and $\delta + 2\eta = \epsilon^2$. There exists a sequence $\{g_n\}$ in C such that $\int |g_n - f|\, d\alpha \longrightarrow 0$, and we may assume that $\int |g_n|\, d\alpha \leq \delta + \eta$ for all n. Lemma 1 implies that $|g_n(x)| \leq \sqrt{\delta + \eta}$ in $X - X_n$, where X_n is $\sqrt{\delta + \eta}$-small. Let $\eta_1 = \epsilon - \sqrt{\delta + \eta} > 0$; there exists a subsequence $\{h_n\}$ of $\{g_n\}$ such that

$$\int |h_{n+1} - h_n|\, d\alpha \leq \left(\frac{\eta_1}{2^{n+2}}\right)^2.$$

Again by Lemma 1, $|h_{n+1}(x) - h_n(x)| \leq 2^{-(n+2)}\eta_1$ in $X - Y_n$ where Y_n is $2^{-(n+2)}\eta_1$-small $(n \geq 2)$. Since $|h_1(x)| \leq \epsilon - \eta_1$ in $X - Y_1$, where Y_1 is $\sqrt{\delta + \eta}$-small, and since $h_n(x)$ converges to $f(x)$ almost everywhere by Proposition 2, it follows that $|f(x)| \leq \epsilon - \frac{1}{2}\eta_1$ almost everywhere in $X - Y$ where $Y = \bigcup_{n=1}^{\infty} Y_n$. It follows that $|f(x)| > \epsilon$ in $X - X_\epsilon$ where X_ϵ is ϵ-small.

PROPOSITION 1': *If* $\{f_n\}$ *is a* $\hat{p}_\alpha$-*Cauchy sequence in* $\mathfrak{L}(\alpha)$, *there exists a subsequence converging a.u. to some* $f \in \mathfrak{L}(\alpha)$.†

Proof: There exist elements $g_n \in C (n \in \mathbf{N})$ such that $\int |f_n - g_n|\, d\alpha \longrightarrow 0$. Then $\{g_n\}$ is a p_α-Cauchy sequence, and by Proposition 1 there exists a subsequence $\{h_n\}$ converging a.u. to some $f \in \mathfrak{L}(\alpha)$. It can now be concluded from Lemma 1' that the corresponding subsequence of $\{f_n\}$ converges a.u. to f.

We establish three basic convergence theorems.

THEOREM 2 (B. Levi): *Let* $\{f_n\}$ *be a monotone sequence in* $\mathfrak{L}(\alpha)$. *Then* $\{f_n\}$ *converges a.u. to some* $f \in \mathfrak{L}(\alpha)$ *if and only if* $\sup_n |\int f_n\, d\alpha| < +\infty$, *and in this case* $\int (f - f_n)\, d\alpha \longrightarrow 0$.

Proof: $\{f_n\}$ is a $\hat{p}_\alpha$-Cauchy sequence if and only if $\{\int f_n\, d\alpha\}$ is a bounded sequence. In the latter case, there exists a subsequence $\{h_n\}$ converging a.u. to some $f \in \mathfrak{L}(\alpha)$, by Proposition 1'. But

† It is evident that $\int |f_n - f|\, d\alpha \longrightarrow 0$.

since $\{f_n\}$ is monotone, we may take $\{f_n\} = \{h_n\}$, and the proof is complete.

THEOREM 3 (Fatou): *Let* $f_n \in \mathfrak{L}(\alpha)$ *and* $f_n \geq g \in \mathfrak{L}(\alpha)$ $(n \in \mathbf{N})$. *If* $\underline{\lim}_n \int f_n \, d\alpha < +\infty$, *then* $\underline{\lim}_n f_n(x)$ *is finite a.e. and if* $h \in \mathbf{R}^X$ *is such that* $h(x) = \underline{\lim}_n f_n(x)$ *a.e., then* $h \in \mathfrak{L}(\alpha)$ *and* $\int h \, d\alpha \leq \underline{\lim}_n \int f_n \, d\alpha$.

Proof: Instead of considering the sequence $\{f_n - g\}$, let us assume that $f_n \geq 0$. For each $n \in \mathbf{N}$ and $p \geq n$, define $h_{n,p} = \inf \{f_\nu \colon \nu \leq s \leq p\}$; then (since $\mathfrak{L}(\alpha)$ is a lattice) $h_{n,p} \in \mathfrak{L}(\alpha)$ and $h_{n,p} \geq h_{n,p+1} \geq 0$. It follows from Theorem 2 that $h_n(x) = \lim_{p \to \infty} h_{n,p}(x)$ defines a summable function h_n, and obviously $\int h_n \, d\alpha \leq \underline{\lim}_{n \to \infty} \int f_n \, d\alpha < +\infty$. On the other hand, $h_n(x) \leq h_{n+1}(x)$ $(x \in X, n \in \mathbf{N})$; hence, again by Theorem 2, h_n converges a.e. to a summable function h with $\int h \, d\alpha = \lim_{n \to \infty} \int h_n \, d\alpha$. Since $h(x) = \underline{\lim}_n f_n(x)$ wherever this lower limit is finite, the theorem is proved.

THEOREM 4 (Lebesgue): *Let* $\{f_n\}$ *be a sequence in* $\mathfrak{L}(\alpha)$ *such that* $|f_n| \leq g (n \in \mathbf{N})$ *for some* $g \in \mathfrak{L}(\alpha)$, *and such that* $\lim_{n \to \infty} f_n(x) = f(x)$ *a.e. for some* $f \in \mathbf{R}^X$. *Then* $f \in \mathfrak{L}(\alpha)$ *and*

$$\int |f_n - f| \, d\alpha \longrightarrow 0$$

(hence, in particular, $\int f_n \, d\alpha \longrightarrow \int f \, d\alpha$).

Proof: It follows from the assumption $|f_n| \leq g \in \mathfrak{L}(\alpha)$ that $\{\int f_n \, d\alpha \colon n \in \mathbf{N}\}$ is a bounded sequence, and that $-g \leq f_n \leq g$ for all n. By Theorem 3, $f \in \mathfrak{L}(\alpha)$ and $\int f \, d\alpha \leq \underline{\lim}_n \int f_n \, d\alpha$; applying Theorem 3 to $\{-f_n\}$, we obtain $-\int f \, d\alpha \leq \underline{\lim}_n (-\int f_n \, d\alpha)$ $= -\overline{\lim}_n \int f_n \, d\alpha$. It follows that

$$\int f \, d\alpha \leq \underline{\lim}_n \int f_n \, d\alpha \leq \overline{\lim}_n \int f_n \, d\alpha \leq \int f \, d\alpha.$$

Hence $\lim_{n\to\infty} \int f_n\,d\alpha$ exists and is equal to $\int f\,d\alpha$. Now the sequence $\{|f_n - f|\}$ converges to 0 a.e., and we have $|f_n - f| \leq (g + |f|) \in \mathfrak{L}(\alpha)$ for all n. Hence by the result we have just established, $\int |f_n - f|\,d\alpha \longrightarrow 0$ and the proof is complete.

Theorem 2 is also known as the "monotone convergence theorem," Theorem 3 as Fatou's lemma, and Theorem 4 as the "dominated convergence theorem." It is easy to see that in the hypothesis of each of these theorems, any relation of the form "$f \leq g$" ($f, g \in \mathfrak{L}(\alpha)$) may be replaced by "$f(x) \leq g(x)$ a.e."

5. MEASURE

Let S be a set. A nonempty family $\mathcal{S}$ of subsets of S is called a *ring* if $A \in \mathcal{S}$, $B \in \mathcal{S}$ implies $A \cup B \in \mathcal{S}$ and $A - B \in \mathcal{S}$. It follows that the empty set $\varnothing \in \mathcal{S}$, and that the symmetric difference

$$A \,\Delta B = (A - B) \cup (B - A)$$

is in $\mathcal{S}$; hence, since $A \cap B = A \cup B - A \,\Delta B$, $A \cap B$ is in $\mathcal{S}$. $\mathcal{S}$ is an *algebra* if $\mathcal{S}$ is a ring and $S \in \mathcal{S}$; then for every $A \in \mathcal{S}$, its complement $A' = S - A$ is in $\mathcal{S}$. A ring $\mathcal{S}$ is a *σ-ring* if $A_n \in \mathcal{S}$ $(n \in \mathbf{N})$ implies $\bigcup_{n=1}^{\infty} A_n \in \mathcal{S}$; since $\bigcap_1^{\infty} A_n = A_1 - \bigcup_1^{\infty} (A_1 - A_n)$, it follows that $A_n \in \mathcal{S}$ $(n \in \mathbf{N})$ implies $\bigcap_1^{\infty} A_n \in \mathcal{S}$ if $\mathcal{S}$ is a σ-ring. A *σ-algebra* is a σ-ring and an algebra.

It is an easy exercise to show that, given a set S, the intersection of any collection of rings (resp. σ-rings, algebras, σ-algebras) of subsets of S is again a ring (resp. σ-ring, algebra, σ-algebra). Hence if $\mathfrak{F}$ is an arbitrary nonempty family of subsets of S, there exists a smallest ring $\mathcal{S}(\mathfrak{F})$ containing the sets of $\mathfrak{F}$, and correspondingly for σ-rings, algebras, and σ-algebras. $\mathcal{S}(\mathfrak{F})$ is called the ring (σ-ring, etc.) *generated* by $\mathfrak{F}$.

If $\mathfrak{F}$ is the family of all intervals in X, the sets in the σ-ring $\mathcal{B}$ generated by $\mathfrak{F}$ are called the *Borel sets* in X. Evidently, $\mathcal{B}$ is a σ-algebra; in particular, every open subset of X (being the union of countably many intervals) is a Borel set, and hence every closed subset of X is a Borel set. Every finite, and hence every countable, subset of X is a Borel set.

DEFINITION 3: *A subset $A \subset X$ is α-measurable if it is the union of* (at most) *countably many disjoint subsets of X whose characteristic functions*† *are α-summable. If $A = \bigcup_1^\infty A_n$ is any such representation, then the α-measure of A is defined as the number* (possibly $+\infty$)

$$m_\alpha(A) = \sum_1^\infty \int \chi_{A_n}\, d\alpha.$$

We shall show immediately that $m_\alpha(A)$ is defined unambiguously (i.e., independently of the particular representation of A). First we prove

PROPOSITION 5: *Every Borel set in X is α-measurable. The family of subsets of X whose characteristic functions are summable forms a ring, and the family $\mathfrak{M}(\alpha)$ of all α-measurable sets is a σ-algebra.*

Proof: Let χ be the characteristic function of an interval $I \subset X$. Obviously $\chi = \lim f_n$ where $\{f_n\}$ is a decreasing sequence of non-negative functions in C. Since $\int f_n\, d\alpha \geq 0$ $(n \in \mathbf{N})$, Theorem 2 implies that $\chi \in \mathfrak{L}(\alpha)$, i.e., I is α-measurable. Further if χ_1, χ_2 are α-summable characteristic functions of two sets A_1, A_2 in X, then $\sup(\chi_1, \chi_2)$ (resp. $\chi_1 - \inf(\chi_1, \chi_2)$) are the characteristic functions of $A_1 \cup A_2$ (resp. $A_1 - A_2$), and are summable since $\mathfrak{L}(\alpha)$ is a vector lattice. It follows now that the family of sets with summable characteristic functions is a ring. By definition of $\mathfrak{M}(\alpha)$, it is established that $\mathfrak{M}(\alpha)$ is a σ-ring containing all Borel sets, and hence a σ-algebra.

A function $A \longrightarrow m(A)$ on a ring $\mathcal{S}$ into the extended interval $[0, \infty]$ is *countably additive* if $m(A) = \sum_1^\infty m(A_n)$ if $\{A_n\}$ is a disjoint family of sets in $\mathcal{S}$ such that $A = \bigcup_1^\infty A_n \in \mathcal{S}$. We show first that

---

† If $A \subset X$, the characteristic function χ_A is the function $\in \mathbf{R}^X$ for which $\chi_A(x) = 1$ if $x \in A$, $\chi_A(x) = 0$ if $x \in X - A$.

$A \longrightarrow \int \chi_A \, d\alpha$ is countably additive on the ring of subsets of X whose characteristic functions are α-summable. If $A = \bigcup_1^\infty A_n$ and A_n are disjoint, then $\chi_A(x) = \sum_1^\infty \chi_{A_n}(x)$, $x \in X$. It follows from Theorem 2 that $\int \chi_A \, d\alpha = \sum_1^\infty \int \chi_{A_n} \, d\alpha$, hence the assertion is true. Now let $A \in \mathfrak{M}(\alpha)$ and let $A = \bigcup_1^\infty A_n = \bigcup_1^\infty A'_n$ be two different representations of A (Definition 3). Set $A_{n,m} = A_n \cap A'_m$, then $A_{n,m}$ are disjoint and $A_n = \bigcup_{m=1}^\infty A_{n,m}$ and $A'_n = \bigcup_{m=1}^\infty A_{m,n}$. By what we have proved,

$$\sum_1^\infty m_\alpha(A'_n) = \sum_{m,n=1}^\infty m_\alpha(A_{n,m}) = \sum_1^\infty m_\alpha(A_n),$$

since a series with nonnegative terms, whether it diverges or not, may be arbitrarily rearranged without changing its sum. This implies that for every $A \in \mathfrak{M}(\alpha)$, $m_\alpha(A)$ is unambiguously defined.

Let $\{A_n\}$ be a sequence of disjoint α-measurable sets, then $A = \bigcup_1^\infty A_n$ is α-measurable by Proposition 5. Let $A_n = \bigcup_{m=1}^\infty A_{n,m}$ be a representation of A_n by means of disjoint sets $A_{n,m}$ whose characteristic functions are summable. Since $\{A_{n,m} : m \in \mathbf{N}, n \in \mathbf{N}\}$ is a disjoint family whose union is A, we obtain

$$m_\alpha(A) = \sum_{m,n=1}^\infty m_\alpha(A_{n,m}) = \sum_{n=1}^\infty \left[\sum_{m=1}^\infty m_\alpha(A_{n,m}) \right] = \sum_{n=1}^\infty m_\alpha(A_n);$$

hence, if we call a countably additive function on a σ-ring (of subsets of some set S) into $[0, \infty]$ a *measure*,† we have proved

PROPOSITION 6: *The set function $A \longrightarrow m_\alpha(A)$ is a measure on the σ-algebra $\mathfrak{M}(\alpha)$.*

† It is clear that a measure m is finitely additive and that $m(\varnothing) = 0$. Also, m is *monotone* and *subtractive*, i.e., if $A \subset B$ and A, B are in the σ-ring in question, then $m(A) \leq m(B)$ and $m(B - A) = m(B) - m(A)$, the latter provided $m(A) < \infty$.

Every interval I is α-measurable (cf. proof of Proposition 5); since χ_I is the limit of a decreasing sequence of functions in C, it follows from Theorem 2 that $m_\alpha(I)$ is finite and, in particular, $m_\alpha(I) \geq \alpha(I)$. Here equality need not hold; if, e.g., $X = \mathbf{R}$ and $\alpha(I) = f(b) - f(a)$ $(I = [a, b])$ where f is an increasing real function (on $\mathbf{R}$) discontinuous at $x = x_0$, then for $I = [x_0, x_0]$ we have $0 = \alpha(I) < m_\alpha(I) = f(x_0 + 0) - f(x_0 - 0)$. Similar remarks are valid for open intervals; if I is an open interval then it is a Borel set, hence measurable, and we have $m_\alpha(I) \leq \alpha(\bar{I})$.

Anticipating Theorem 5, we may now express the notions "ϵ-small $[\alpha]$" and "zero set $[\alpha]$" in terms of measure. By Definition 1, A is ϵ-small if there exists a collection of intervals $\{I_n\}$ with $\sum_1^\infty \alpha(I_n) \leq \epsilon$ and $A \subset \bigcup_1^\infty J_n$ where J_n denotes the interior of $\bigcup_{\nu=1}^n I_\nu$; since $m_\alpha(J_n) \leq \sum_1^n \alpha(I_\nu)$, every ϵ-small set is a subset of an open set with measure $\leq \epsilon$. Conversely, by Theorem 5 it is easily established that every set A with $m_\alpha(A) < \epsilon$ $(\epsilon > 0)$ is ϵ-small. A is a zero set $[\alpha]$ if and only if $m_\alpha(A) = 0$.

A measure m on a σ-ring is *finite* (resp. *σ-finite*) if every measurable set is of finite measure (resp. a countable union of sets of finite measure); if, in addition, S is measurable (i.e., if $\mathcal{S}$ is an algebra), then m is called *totally finite* (resp. *totally σ-finite*). Each m_α on X is σ-finite by definition and totally σ-finite since X is a Borel set. However, $m_\alpha(X) = \infty$ in general.

What is the relation between the algebra $\mathcal{B}$ of Borel sets and $\mathfrak{M}(\alpha)$? Evidently, $\mathcal{B}$ is a subalgebra of each $\mathfrak{M}(\alpha)$. We shall see that the Borel sets of X are "dense" in $\mathfrak{M}(\alpha)$ in a certain sense; more precisely, it follows from Theorem 5 that $\mathfrak{M}(\alpha)$ is the σ-algebra generated by the family of all Borel sets and all sets of α-measure 0.

THEOREM 5: *Let A be an α-measurable set of finite measure. There exists, for each $\epsilon > 0$, an open set G_ϵ and a closed set F_ϵ such that $F_\epsilon \subset A \subset G_\epsilon$ and $m_\alpha(G_\epsilon - F_\epsilon) \leq \epsilon$. If A is an arbitrary α-measurable set, there exists a Borel set H such that $m_\alpha(A \,\Delta H) = 0$.*

Proof: Let $m_\alpha(A) < +\infty$ and η be a positive number $< \frac{1}{2}$. If χ_A is the characteristic function of A, χ_A is summable; by Theorem

1 there exists an $f \in C$ (i.e., a continuous f with compact support) such that $\int |f - \chi_A| \, d\alpha < \eta^2$. From Lemma 1′ (Sec. 4) it follows that $|f(x) - \chi_A(x)| < \eta$ in $X - X_\eta$, where X_η is η-small. By Definition 1, X_η may be assumed as an open set such that $m_\alpha(X_\eta) \leq \eta$. $Y = \{x: |f(x)| > \eta\}$ is an open set in X. We have

$$Y - X_\eta \subset A \subset Y \cup X_\eta;$$

for if $x \in Y - X_\eta$, then $|f(x)| > \eta$ and $|f(x) - \chi_A(x)| < \eta$ imply $\chi_A(x) \neq 0$, i.e., $x \in A$. The second inclusion is similarly proved. It follows from the additivity of m_α that

$$m_\alpha(Y) - m_\alpha(X_\eta) \leq m_\alpha(A) \leq m_\alpha(Y) + m_\alpha(X_\eta).$$

It follows then from the last inequalities that $m_\alpha(G - A) \leq 2\eta$ where $A \subset G$ and $G = Y \cup X_\eta$ is an open set. Next we show that $m_\alpha(A - F)$ is arbitrarily small for a suitable closed set F. We observe that with $Z = \{x: |f(x)| \geq \eta\}$, which is a closed set, we still have

$$Z - X_\eta \subset A \subset Z \cup X_\eta;$$

but $Z - X_\eta$ (which is the intersection of Z with the complement $X - X_\eta$) is closed if X_η is open. It follows again that $m_\alpha(F - A) \leq 2\eta$ with $F = Z - X_\eta$. Hence $m_\alpha(G - F) \leq \epsilon$ if $\eta \leq \epsilon/4$, and the first part of the theorem is established.

Now let U_n (resp. F_n) be a sequence of open (resp. closed) sets such that $F_n \subset A \subset U_n$ and $m_\alpha(U_n - F_n) \leq 1/n$. If $U = \bigcap_1^\infty U_n$ and $F = \bigcup_1^\infty F_n$, then U and F are Borel sets with $F \subset A \subset U$ and $m_\alpha(U - F) = 0$. Setting $H = U$ (or F), $A \,\Delta H$ is a subset of $U - F$ and hence $m_\alpha(A \,\Delta H) = 0$. Finally, if A is an arbitrary measurable set, then $A = \bigcup_1^\infty A_n$ where $m_\alpha(A_n) < +\infty$ (Definition 3). There exist Borel sets H_n and K_n ($n \in \mathbf{N}$) with $H_n \subset A_n \subset K_n$ and such that $m_\alpha(K_n - H_n) = 0$. If $H = \bigcup_1^\infty H_n$, $K = \bigcup_1^\infty K_n$, then $H \subset A \subset K$, and $m_\alpha(A - H) = m_\alpha(K - A) = 0$ since a countable union of zero sets is a zero set. This completes the proof.

REMARK: The preceding proof implies that for a set of finite measure, the Borel set H may be chosen as a $G_\delta \supset A$ or an $F_\sigma \subset A$. (A G_δ (resp. F_σ) is a countable intersection (resp. union) of open (resp. closed) subsets of X.)

We shall point out briefly how m_α determines a measure, in a natural way, on certain σ-algebras. Let A be a fixed α-measurable set in X; a real function $f \in \mathbf{R}^A$ is called α-summable *over* A if the function f^* for which $f^*(x) = f(x)$ $(x \in A)$ and $f^*(x) = 0$ $(x \in X - A)$, is in $\mathfrak{L}(\alpha)$. The set of all these f^* forms obviously a vector sublattice $\mathfrak{L}(\alpha, A)$ of $\mathfrak{L}(\alpha)$; the integral on A, defined as

$$\int_A f\, d\alpha = \int f^*\, d\alpha$$

is the restriction of $f \longrightarrow \int f\, d\alpha$ to $\mathfrak{L}(\alpha, A)$. It is easy to see that the corresponding subset $L(\alpha, A)$ of $L(\alpha)$ (i.e., the set of equivalence classes containing some f^*) is a sublattice and a closed vector subspace of $L(\alpha)$; hence $L(\alpha, A)$ is a Banach lattice. $L(\alpha, A)$ is referred to as the Banach space of (equivalence classes mod. null functions of) α-summable real functions over A. It is not hard to prove that $L(\alpha, A) + L(\alpha, X - A) = L(\alpha)$ is a direct sum decomposition of $L(\alpha)$. By the method used in Definition 3, the integral $\int_A |f|\, d\alpha$ determines a measure on the family $\mathfrak{M}(\alpha, A) = \{B \cap A : B \in \mathfrak{M}(\alpha)\}$ of subsets of A, which is a σ-algebra.

The preceding discussion leads to the following concept. An $f \in \mathbf{R}^X$ is *locally α-summable* if the restriction of f to every bounded α-measurable set A is α-summable over A; thus every continuous function on X is locally summable for every α. The integral $A \longrightarrow \int_A f\, d\alpha$, for fixed locally summable $f \geq 0$, is a countably additive function on the ring of bounded α-measurable subsets of X; it is not hard to show that this generates a measure $m_{f,\alpha}$ on $\mathfrak{M}(\alpha)$ (cf. Definition 3), but we shall not pursue these considerations any further. If f is continuous, it should be noted (in analogy to the remark made subsequent to Proposition 6) that the measure so defined is not in general an extension of the additive function of intervals $I \longrightarrow \int_I f\,d\alpha$ described in Sec. 2 (where $\int_I f\,d\alpha$ denotes the Riemann-Stieltjes integral over I).

6. MEASURABLE FUNCTIONS

DEFINITION 4: *A function $f \in \mathbf{R}^X$ is a Borel* (resp. *α-measurable*) *function if for every Borel set $M \subset \mathbf{R}$, $f^{-1}(M)$ is a Borel* (resp. *α-measurable*) *subset of X.*

It follows from this definition that for every α-measurable set A, the characteristic function χ_A is α-measurable; for if M is a Borel set containing none or both of the numbers 0, 1, then $f^{-1}(M) = \emptyset$ or $f^{-1}(M) = X$; if M contains exactly one of them, then either $f^{-1}(M) = A$ or $X - A$. Further, if f is α-measurable and $f(x) = g(x)$ a.e., then g is α-measurable; for $f^{-1}(M)\,\Delta g^{-1}(M) = N$ where N is an α-zero set. Another remark is of importance for some of the proofs that follow: If f is a mapping on a set S_1 into a set S_2, and $\mathcal{S}$ is a ring (resp. σ-ring, algebra, σ-algebra) in S_1, then $\{S \subset S_2 : f^{-1}(S) \in \mathcal{S}\}$ is also a ring (resp. σ-ring, algebra, σ-algebra). The proof is easy and will be omitted. Hence, since the family of all Borel sets in X is generated by all open sets (every open set is the union of countable many intervals, and every interval is a G_δ), in order to show that $f \in \mathbf{R}^X$ is Borel (resp. α-measurable), it suffices to show that $f^{-1}(G)$ is a Borel (resp. α-measurable) set for every open set, or, more generally, for every set of a generating family of $\mathcal{B}(\mathbf{R})$.

PROPOSITION 7: *Every continuous real function on X is a Borel function* (hence α-measurable for every α).

Proof: Since $f^{-1}(G)$ is an open subset of X for every open $G \subset \mathbf{R}$, it follows from the foregoing remarks that f is a Borel function.

PROPOSITION 8: *If $f_1, f_2, \cdots$ are Borel* (resp. *α-measurable*) *functions, and if $f_0(x) = \lim_{n\to\infty} f_n(x)$ everywhere* (resp. *a.e.*) *in X, then f_0 is a Borel* (resp. *α-measurable*) *function.*

Proof: Since α-measurability of a function is not affected if its values are changed in an α-zero set, we may suppose

$f_n(x) \longrightarrow f_0(x)$ $(x \in X)$ in both cases. If we set $A = \{x: c_1 < f_0(x) < c_2\}$ and

$$A_m = \bigcap_{n=m}^{\infty} \left\{x: c_1 + \frac{1}{m} < f_n(x) < c_2 - \frac{1}{m}\right\}$$

for any two real numbers, then it follows from the assumptions on $\{f_n\}$ that A_m is a Borel (resp. α-measurable) set. Since $A = \bigcup_1^{\infty} A_m$, the assertion follows from the remark preceding Proposition 7.

PROPOSITION 9: *Let* $\{f_\mu: \mu = 1, \cdots, m\}$ *be Borel* (resp. *α-measurable) functions on* X, *and let* g *be a Borel function on* $\mathbf{R}^m$ *into* $\mathbf{R}$. *Then the function* $h \in \mathbf{R}^X$ *defined by*

$$h(x) = g(f_1(x), f_2(x), \cdots, f_m(x))$$

is a Borel function (resp. *α-measurable*) *on* X.

Proof: Since h is the composite map $g \circ \psi$, where ψ is the mapping $x \longrightarrow (f_1(x), \cdots, f_m(x))$ on X into $\mathbf{R}^m$, and since the family of all open intervals in $\mathbf{R}^m$ generates the σ-ring of Borel sets in $\mathbf{R}^m$, the proposition will be proved if we show that for every open interval $M \subset \mathbf{R}^m$, $\psi^{-1}(M)$ is a Borel (resp. α-measurable) set in X. If we denote by M_μ the projection of M onto the μth coordinate, then M_μ is an open interval in $\mathbf{R}$. On the other hand, $\psi^{-1}(M) = \bigcap_{\mu=1}^{m} f_\mu^{-1}(M_\mu)$; hence the assertion is proved.

COROLLARY: *If* g *is continuous on* $\mathbf{R}^m$ *(into* $\mathbf{R}$*), and* $\{f_\mu: \mu = 1, \cdots, m\}$ *are Borel* (resp. *α-measurable) functions on* X, *then* $x \longrightarrow g(f_1(x), f_2(x), \cdots, f_m(x))$ *is a Borel* (resp. *α-measurable*) *function on* X.

THEOREM 6: *The family of all* (finite) *Borel* (resp. *α-measurable*) *functions on* X *forms a vector sublattice of* $\mathbf{R}^X$ *which is a subalgebra*† *and is closed under the formation of pointwise* (resp. *a.e. pointwise*) *sequential limits.*

Proof: Since $(x, y) \longrightarrow x + y$, $(x, y) \longrightarrow xy$, $(x, y) \longrightarrow \sup(x, y)$, and $(x, y) \longrightarrow \inf(x, y)$ are all continuous on $\mathbf{R}^2$ into $\mathbf{R}$, it follows

† $\mathbf{R}^X$ is an algebra with respect to its (natural) vector space structure and the multiplication $(f, g) \longrightarrow fg$, where $fg(x) = f(x)g(x)$ $(x \in X)$.

from the preceding corollary that the families in question form a sublattice and a subalgebra of $\mathbf{R}^X$. The remainder of the assertion is contained in Proposition 8.

REMARK: It follows from Theorem 6 that if $\{f_n\}$ are Borel (resp. α-measurable) functions and $\sup \{f_n(x): n \in \mathbf{N}\} < +\infty$ $(x \in X)$, then $\sup f_n$ is Borel (resp. α-measurable) since $\sup f_n = \lim_{n\to\infty} \{ \sup_{1 \le \nu \le n} f_\nu\}$. Similar remarks apply to inf, $\overline{\lim}$, and $\underline{\lim}$.

Our further purpose in this section is to study the relationship between the Borel functions, the α-summable functions, and the α-measurable functions on X. We have seen that every characteristic function χ_A of an α-measurable set A is α-measurable; then so is $x \longrightarrow \sum_{i=1}^{n} c_i \chi_{A_i}(x)$ $(c_i \in \mathbf{R})$, if A_i are α-measurable $(i = 1, \cdots, n)$. If the A_i are disjoint, such a function is called a *simple function.* We have

PROPOSITION 10: *Every α-measurable* (resp. *nonnegative α-measurable*) *function is the limit of a sequence of* (resp. *increasing sequence of nonnegative*) *simple functions. A corresponding statement holds for Borel functions.*†

Proof: If $f^+ = \sup(f, 0)$, $f^- = \sup(-f, 0)$ then $f = f^+ - f^-$ and f^+, f^- are α-measurable by Theorem 6. Hence it is sufficient to show that $f_n(x) \longrightarrow f^+(x)$ for an increasing sequence of simple functions $f_n \ge 0$. Now define

$$f_n(x) = \begin{cases} \dfrac{\nu - 1}{2^n} & \text{if } \dfrac{\nu - 1}{2^n} \le f^+(x) < \dfrac{\nu}{2^n}\ \nu = 1, \cdots, n2^n. \\ n & \text{if } f^+(x) \ge n. \end{cases}$$

Then the sequence $\{f_n\}$ satisfies the requirement.

THEOREM 7: *In order that $f \in \mathbf{R}^X$ be α-summable, it is necessary and sufficient that f be α-measurable, and that $|f|$ be α-summable. If f is α-measurable and $|f| \le g$ for some $g \in \mathfrak{L}(\alpha)$, then f is α-summable.*

---

† With the sequence consisting of simple Borel functions. (A simple function f is Borel if and only if the A_i, in some representation of f, are Borel sets.)

Proof: If $f \in \mathbf{R}^X$ is summable, then $|f|$ is α-summable since $\mathfrak{L}(\alpha)$ is a lattice; since, by definition, f is the a.u. limit of a sequence of continuous (hence α-measurable, Proposition 7) functions, f is α-measurable by Proposition 8. On the other hand, if $|f|$ is α-summable and f is α-measurable, then because of $f = |f| - 2f^-$, f is α-summable if f^- is. Let $0 \leq f_n \uparrow f^-$ be a sequence of simple functions. If each f_n is α-summable, then so is f^- by Theorem 2; for then the integrals $\int f_n\, d\alpha$ are uniformly bounded ($\leq \int |f|\, d\alpha$). All that remains to prove is that if h is a simple function with $0 \leq h \leq f$, where f is α-summable, then $h \in \mathfrak{L}(\alpha)$. Let $h = \sum_{i=1}^{n} c_i \chi_{A_i} (c_i > 0)$ be a representation of h, and let $A_i = \bigcup_{n=1}^{\infty} A_{i,n}$ where $A_{i,n}$ are disjoint sets whose characteristic functions are α-summable (Definition 3). Since $\chi_{A_i} = \sum_{n=1}^{\infty} \chi_{A_{i,n}}$ and $\chi_{A_i} \leq (c_i)^{-1} f$, it follows from Theorem 2 that χ_{A_i}, hence h, is α-summable. We have actually shown that if f is measurable and $|f| \leq g$ for some $g \in \mathfrak{L}(\alpha)$, then $f \in \mathfrak{L}(\alpha)$, and the proof is complete.

We say that a function $f \in \mathbf{R}^A$ is α-essentially bounded on $A \subset X$ if f is bounded on $A - N$ where N is an α-zero set. The essential (absolute) supremum is defined as the g.l.b. of all $c > 0$ such that $m_\alpha\{x: |f(x)| > c\} = 0$, and denoted sup ess $|f|$ or, more precisely, sup ess $\{|f(x)|: x \in A\}$. If A is an α-measurable set, $f \in \mathbf{R}^A$ is α-measurable on A if for every Borel set $M \subset \mathbf{R}$, $f^{-1}(M)$ is an α-measurable subset of A or, equivalently, if $(f^*)^{-1}(M)$ is α-measurable where $f^*(x) = f(x)$ in A and $f^*(x) = 0$ in $X - A$ (cf. Sec. 5). With this notation, we obtain

PROPOSITION 11: *If $f \in \mathbf{R}^A$ is an α-essentially bounded function, measurable over a set A of finite α-measure, then f is α-summable over A.*

Proof: We have to show that f^* is α-summable. If $|f(x)| \leq c$ a.e. in A, then $|f^*(x)| \leq c\chi_A(x)$ a.e. Hence, by Theorem 7, f^* is α-summable.

It follows, in particular, from Proposition 11 that an α-measurable f essentially bounded on bounded subsets of X is locally α-summable (Sec. 5).

We establish now a relationship between Borel functions and α-measurable functions.

THEOREM 8: *For every α-measurable function g, there exists a Borel function f such that $f(x) = g(x)$ a.e. $[\alpha]$.*

Proof: It is evidently sufficient to prove the statement for $g \geq 0$. By Proposition 10, g is the limit of an increasing sequence of simple functions; hence if we show that for every simple function $h \geq 0$, there exists a (simple) Borel function f such that $0 \leq f \leq h$ and $h(x) = f(x)$ a.e. $[\alpha]$, the proof will be complete by Proposition 8. Now let $h \geq 0$ be a simple function with representation $h = \sum_{i=1}^{n} c_i \chi_{A_i}$, where A_i $(i = 1, \cdots, n)$ are α-measurable disjoint sets. There exist, by the remark subsequent to Theorem 5, Borel sets $F_i \subset A_i$ with $m_\alpha(A_i - F_i) = 0$; then $f = \sum_{i=1}^{n} c_i \chi_{F_i}$ is a Borel function such that $0 \leq f \leq h$ and $h(x) = f(x)$ a.e.

Reviewing our results, we have $C \subset \mathfrak{L}(\alpha) \subset \mathfrak{M}(\alpha) \subset \mathbf{R}^X$, if by $\mathfrak{M}(\alpha)$ we denote the family of α-measurable functions; and from the preceding theorems one may prove that $C \neq \mathfrak{L}(\alpha)$, and $\mathfrak{L}(\alpha) \neq \mathfrak{M}(\alpha)$ unless m_α vanishes on all but a finite number of disjoint α-measurable sets. Also, $\mathfrak{M}(\alpha) \neq \mathbf{R}^X$ in general as we cannot show here. On the other hand, C is dense in $\mathfrak{L}(\alpha)$ with respect to the "seminorm" $\hat{p}_\alpha$ (Sec. 3), while $\mathfrak{L}(\alpha)$ may be characterized as those $f \in \mathfrak{M}(\alpha)$ for which $|f|$ has a "finite integral" (i.e., for which the measure $m_{|f|,\alpha}$ mentioned at the end of Sec. 5 is totally finite). As for the family $\mathfrak{B}$ of Borel functions, $\mathfrak{B}$ is "dense" in $\mathfrak{M}(\alpha)$ and $\mathfrak{L}(\alpha)$ in the sense made precise by Theorem 8, and contains C. It follows from this that every class in the Banach space $L(\alpha)$ contains at least one Borel function; thus, had we started with $\mathfrak{B}$ in place of $\mathbf{R}^X$, we would have obtained the same spaces $L(\alpha)$ as quotient spaces derived from one and the same space $\mathfrak{B}$, independently of α.

We remark further that it is customary in real analysis to work with the set $(\mathbf{R}^*)^X$, where $\mathbf{R}^*$ is the extended real-number system. It is true that for $f, g \in (\mathbf{R}^*)^X$ and $c \in \mathbf{R}^*$, $f + g$, cf, and fg can be consistently defined as elements of $(\mathbf{R}^*)^X$, but the vector space properties of $\mathbf{R}^X$ cannot be saved. For the definition of Borel (and α-measurable) functions, Definition 4 then has to be changed so as to admit $\{+\infty\}$ and $\{-\infty\}$ as additional "Borel sets." Compared to this more general approach, we have lost little at least with respect to the spaces $\mathfrak{L}(\alpha)$ (it can be shown that every α-summable $f \in (\mathbf{R}^*)^X$ takes the values $\pm\infty$ in a zero set $[\alpha]$). Neither is the loss with respect to $\mathfrak{M}(\alpha)$ considerable; the extended α-measurable functions may indeed be thought of as pointwise limits of sequences in $\mathfrak{M}(\alpha)$. On the other hand, working in $\mathbf{R}^X$ permits an unobliterated discussion of the algebraic and order structures involved, and the spaces $L(\alpha)$ are entirely unaffected.

We remark in conclusion that the subset of $\mathfrak{M}(\alpha)$ for whose elements $|f|^p$ ($p \geq 1$, fixed) is summable, forms a vector sublattice $\mathfrak{L}_p(\alpha)$ of $\mathfrak{M}(\alpha)$; with respect to the norm

$$[f] \longrightarrow \left(\int |f|^p \, d\alpha\right)^{1/p},$$

the quotient space of $\mathfrak{L}_p(\alpha)$ (modulo null functions) is a Banach lattice $L_p(\alpha)$. Similarly, the quotient space of essentially bounded α-measurable functions is a Banach lattice $M(\alpha)$ with respect to the norm $[f] \longrightarrow \sup \operatorname{ess} |f|$.

7. PRODUCT MEASURE; THE FUBINI THEOREM

We shall assume throughout this section that $X = \mathbf{R}^r$ for some fixed $r \geq 2$, and we shall think of a fixed factorization $X = X_1 \times X_2$ where $X_1 = \mathbf{R}^{r_1}$, $X_2 = \mathbf{R}^{r_2}$ such that $r = r_1 + r_2$ ($r_1, r_2 \geq 1$). If $I = I' \times I''$ is the (unique) representation of an interval as the product of its "sides" $I' \subset X_1$ and $I'' \subset X_2$, we shall consider only (nonnegative) finitely additive functions α for which

$$\alpha(I) = \alpha_1(I')\alpha_2(I''),$$

where α_1 (resp. α_2) is a function of the same type on intervals in X_1 (resp. X_2). It is clear from Sec. 5 that each of the functions α, α_1, α_2 gives rise to a (Lebesgue-Stieltjes) measure m_α, m_{α_1}, m_{α_2}, and it turns out that for every interval I, we have

$$m_\alpha(I) = m_{\alpha_1}(I')m_{\alpha_2}(I'');$$

it is for this reason that m_α may justifiably be called a *product measure.* We shall restrict attention in this section to the proof of an important result, known as the Fubini theorem.

If $A \subset X$ is a set, then we shall call the subset

$$A_{x_1} = \{x_2: (x_1, x_2) \in A\} \subset X_2$$

an x_1-section (more precisely, the section determined by x_1) of A; A_{x_2} is defined similarly as a subset of X_1. Accordingly, if $f \in \mathbf{R}^A$, f_{x_1} will denote the restriction to A_{x_1} of f. For convenience of expression, an α-summable function f is said to have property (F) if the following statement is true:

(F) *There is an α_1-zero set $N_1 \subset X_1$ such that for $x_1 \in X_1 - N_1$, f_{x_1} is α_2-summable, and if $g \in \mathbf{R}^{X_1}$ is such that $g(x_1) = \int f_{x_1}\, d\alpha_2$ outside of an α_1-zero set containing N_1, then g is α_1-summable and*

$$\int g\, d\alpha_1 = \int f\, d\alpha.$$

It is not difficult to establish that every $f \in C$ has property (F); i.e., if f is continuous on X and of compact support, then

$$\int f\, d\alpha = \int \left[\int f(x_1, x_2)\, d\alpha_2 \right] d\alpha_1;$$

here N_1 may be taken as the empty set. We omit the proof.

LEMMA 2: *The class of α-summable functions with property* (F) *is closed under the formation of monotone sequential limits.*

Proof: Let f be α-summable and let $\{f_n\}$ be an increasing sequence of α-summable functions with property (F) such that $\lim_{n\to\infty} f_n(x) = f(x)$ $(x \in X)$. Let $g_n \in \mathbf{R}^{X_1}$ be such that $g_n(x_1) = \int f_n(x_1, x_2)\, d\alpha_2$ for whatever values of x_1 the function $x_2 \longrightarrow f_n(x_1, x_2)$ is α_2-summable; then, by assumption, g_n is α_1-summable and $\int g_n\, d\alpha_1 = \int f_n\, d\alpha$. Since $\int f_n\, d\alpha \leq \int f\, d\alpha$ $(n \in \mathbf{N})$, it follows from Theorem 2 (as $\{g_n\}$ may be assumed increasing) that $\sup g_n(x_1) < +\infty$ if $x_1 \in X_1 - N_1$ where N_1 is an α_1-zero set. Let $g \in \mathbf{R}^{X_1}$ be a function such that $g(x_1) = \sup_n g_n(x_1)$ if $x_1 \in X_1 - N_1$, then $g \in \mathfrak{L}(\alpha_1)$ and $\int g\, d\alpha_1 = \lim_{n\to\infty} \int g_n\, d\alpha_1 = \int f\, d\alpha$. Now let $x_1 \in X_1 - N_1$ be fixed. Because $f_n(x) \longrightarrow f(x)$ for all $x \in X$ and because $\{g_n(x_1)\}$ is a bounded sequence, Theorem 2 implies that $x_2 \longrightarrow f(x_1, x_2)$ is α_2-summable, and that $g(x_1) = \int f(x_1, x_2)\, d\alpha_2$. By the properties of g already established, it follows that f has property (F).

THEOREM 9 (Fubini): *Every α-summable function has property* (F).

Proof: We prove first that every α-summable characteristic function χ_A has property (F). Since the characteristic function of every bounded open (resp. compact) subset of X is the limit of an increasing (resp. decreasing) sequence of functions in C, it follows by Lemma 2 and the assertion stated above that those characteristic functions have (F). Again by Lemma 2, this is true of the characteristic function of every G_δ and F_σ of finite measure, and by the remark subsequent to Theorem 5, there exist an F_σ and a G_δ, such that $F_\sigma \subset A \subset G_\delta$ and $m_\alpha(G_\delta - F_\sigma) = 0$. For the corresponding characteristic functions we have $\chi_F \leq \chi_A \leq \chi_G$, and since χ_F and χ_G have (F), so does χ_A.

Now every nonnegative α-summable function is, by Proposition 10, the monotone limit of a sequence of (summable) simple functions, hence by Lemma 2 has (F); as every α-summable function f is the difference of its positive and negative parts f^+ and f^- (which are summable since $\mathfrak{L}(\alpha)$ is a lattice), the proof is complete.

We state the following two corollaries of the Fubini theorem.

COROLLARY 1: *If $I = I' \times I''$ is an interval in X then*

$$m_\alpha(I) = m_{\alpha_1}(I')m_{\alpha_2}(I'').$$

COROLLARY 2: *Almost every x_1-section of an α-measurable set A is α_2-measurable; in particular, almost every x_1-section of an α-zero set A is an α_2-zero set.*

REFERENCES

1. Birkhoff, G., *Lattice Theory.* New York, A.M.S. Colloq. Publ., 1948.
2. Bourbaki, N., *Elém. de Math.*, Livre VI: *Intégration.* Paris: Hermann, 1952 *et seq.*
3. Graves, L. M., *Theory of Functions of a Real Variable.* New York: McGraw-Hill Book Co., 1956.

4. Halmos, P. R., *Measure Theory.* Princeton, N.J.: Van Nostrand, 1950.
5. Hille, E., *Semi-Groups and Functional Analysis.* New York: A.M.S. Colloq. Publ., 1948.
6. Kelley, J. L., *General Topology.* Princeton, N.J.: Van Nostrand, 1955.
7. Rogosinski, W., *Volume and Integral.* New York: Interscience Publishers, 1952.
8. Saks, S., *Theory of the Integral.* New York: G. E. Stechert & Co., 1937.
9. McShane, E. J., *Integration.* Princeton, N.J.: Princeton University Press, 1944.
10. McShane, E. J., and T. Botts, *Real Analysis.* Princeton, N.J.: Van Nostrand, 1959.
11. Zaanen, A. C., *An Introduction to the Theory of Integration.* Amsterdam: North Holland Publishing Co., 1961.

HARMONIC ANALYSIS

Guido Weiss

1. INTRODUCTORY REMARKS CONCERNING FOURIER SERIES

Classical harmonic analysis deals mainly with the study of Fourier series and integrals. It occupies a central position in that branch of mathematics known as analysis; in fact, it has been described [10, p. xi] as "the meeting ground" of the theory of functions of a real variable and that of analytic functions of a complex variable. Consequently it arises, in a natural way, in several different contexts. Moreover, many basic notions and results in mathematics have been developed by mathematicians working in harmonic analysis. The modern concept of function was first introduced by Dirichlet while studying the convergence of Fourier series; more recently, the theory of distributions (generalized functions) was developed in close connection with the study of Fourier transforms. The Riemann and, later, the Lebesgue integrals were originally introduced in works dealing with harmonic analysis. Infinite cardinal and ordinal numbers, probably the most original and striking notions of modern mathematics, were developed by Cantor in his attempts to solve a delicate real-variable problem involving trigonometric series.

It is our purpose to present some of the main aspects of classical and, to a lesser extent, modern harmonic analysis. The development of the former uses principally the theories of functions of a real variable and of a complex variable while the latter draws heavily from the ideas of abstract functional analysis. Consequently, we shall assume the reader to be acquainted with the material usually presented in a first course in Lebesgue integration or measure theory and to have an elementary knowledge of analytic function theory; moreover, we will require a minimal knowledge of functional analysis.†

Suppose that we are given a real- or complex-valued function f, defined on the real line, periodic of period 1 (that is, $f(x) = f(x+1)$ for all x) and (Lebesgue) integrable when restricted to the interval $(0, 1)$.‡ Its *Fourier transform* is then the function $\hat{f}$, defined on the integers, whose value at k (the kth *Fourier coefficient*) is

$$\hat{f}(k) = \int_0^1 f(t)e^{-2\pi ikt}\,dt, \qquad k = 0, \pm 1, \pm 2, \pm 3, \cdots.$$

The *Fourier series* of f is the series

$$\sum_{k=-\infty}^{\infty} \hat{f}(k)e^{2\pi ikx} \tag{1.1}$$

considered as the sequence of (*symmetric*) *partial sums*

$$s_n(x) = \sum_{k=-n}^{n} \hat{f}(k)e^{2\pi ikx}. \tag{1.2}$$

The reader is undoubtedly familiar with these notions; however, he has probably been introduced to them in a slightly different manner. For example, in classical treatments of Fourier series

---

† The reader should be acquainted with the elementary properties of Banach spaces and Hilbert spaces, as well as those of linear operators acting on these spaces. He is well advised to glance over the article on functional analysis in Volume 1 of this series.

‡ It follows that f must be integrable over any finite subinterval of $(-\infty, \infty)$. In the sequel we shall use the term "periodic" to mean periodic and of period 1.

the term "Fourier transform" is not used. Generally, the sequence of Fourier coefficients

$$c_k = \int_0^1 f(t)e^{-2\pi ikt}\,dt, \qquad k = 0, \pm 1, \pm 2, \pm 3, \cdots,$$

is introduced without emphasizing that one has really defined a function on the integers (the Fourier transform) and the Fourier series of f is usually denoted, simply, by the series $\sum_{k=-\infty}^{\infty} c_k e^{2\pi ikx}$. The above notation and emphasis (as will shortly become apparent) are useful in order to give a unified approach to the theory of Fourier series, Fourier integrals, and their many analogs and extensions. Furthermore, when f is real-valued, the Fourier series of f is often introduced as the series

$$\frac{a_0}{2} + \sum_{k=1}^{\infty} (a_k \cos 2\pi kx + b_k \sin 2\pi kx) \tag{1.1'}$$

considered as the sequence of partial sums

$$s_n(x) = \frac{a_0}{2} + \sum_{k=1}^{n} (a_k \cos 2\pi kx + b_k \sin 2\pi kx), \tag{1.2'}$$

where

$$a_k = 2\int_0^1 f(t) \cos 2\pi kt\,dt \qquad \text{and} \qquad b_k = 2\int_0^1 f(t) \sin 2\pi kt\,dt,$$

$k = 0, 1, 2, 3, \cdots$. It is easy to verify that the two expressions (1.2) and (1.2′) are equal.

In most advanced calculus courses it is shown that the Fourier series (1.1) (or (1.1′)) converges to $f(x)$ provided f is sufficiently well-behaved at the point x. For example, as we shall show in Sec. 3, this occurs whenever f is differentiable at the point x. This is a solution of a very special case of the central problem in the classical study of Fourier series: to determine whether, and in what sense, the series (1.1) represents the function f. Perhaps one of the best ways of penetrating into the subject of harmonic analysis is by studying this problem.

If we pose this problem in the most obvious way by asking if the series (1.1) converges to $f(x)$ for all x or for almost all x (a much more reasonable question, since altering f on a set of measure zero does not alter the Fourier series), we immediately encounter some serious difficulties. In fact, Kolmogoroff [10, p. 310] has shown that there exists a periodic function, integrable on (0, 1), whose Fourier series diverges everywhere. In general, one must impose fairly strong conditions on f in order to obtain the convergence of its Fourier series. Perhaps the most important unsolved problem in the classical theory is the following: does there exist a continuous periodic function whose Fourier series diverges on a set of positive measure? (See note on page 178.)

On the other hand, if we consider only functions in $L^2(0, 1)$, that is, periodic functions f such that

$$\|f\|_2 = \left(\int_0^1 |f(x)|^2\, dx\right)^{1/2} < \infty,$$

we obtain a complete and elegant solution to the central problem announced above restricted to this space and its norm. More precisely, we shall show that in this case the partial sums (1.2) converge to f in the L^2-norm; that is,

$$\lim_{n\to\infty} \|f - s_n\|_2 = \lim_{n\to\infty} \left(\int_0^1 |s_n(x) - f(x)|^2\, dx\right)^{1/2} = 0.$$

This is an immediate consequence of the fact that, with respect to the inner product $(f, g) = \int_0^1 f(x)\overline{g(x)}\, dx$, $L^2(0, 1)$ is "essentially" a Hilbert space† and the exponential functions e_k, $k = 0, \pm 1, \pm 2, \pm 3, \cdots$, where $e_k(x) = e^{2\pi ikx}$, form an *orthonormal basis*; that is,

$$(e_k, e_j) = \int_0^1 e^{2\pi ikx} e^{-2\pi ijx}\, dx = \delta_{kj}, \tag{1.3}$$

where δ_{kj} is the "Kronecker δ" ($\delta_{kj} = 0$ when $k \neq j$ and $\delta_{kk} = 1$), and for each $f \in L^2(0, 1)$

$$\begin{aligned} \lim_{n\to\infty} \left\| \sum_{k=-n}^{n} c_k e_k - f \right\|_2^2 \\ = \lim_{n\to\infty} \left(\sum_{k=-n}^{n} c_k e_k - f,\ \sum_{k=-n}^{n} c_k e_k - f \right) = 0, \end{aligned} \tag{1.4}$$

† More precisely, it is the collection of equivalence classes we obtain by identifying functions in $L^2(0, 1)$ that are equal almost everywhere that forms a Hilbert space.

when $c_k = (f, e_k)$ is the kth Fourier coefficient of f, $k = 0, \pm 1, \pm 2, \cdots$. While the orthogonality relations (1.3) are obvious, the convergence (1.4) will require some proof. A very simple argument, however, shows that the partial sums $s_n = \sum_{k=-n}^{n} c_k e_k = \sum_{k=-n}^{n} \hat{f}(k) e_k$ converge in the L^2-norm to some function in L^2. By the Riesz-Fisher theorem, which asserts that all the L^p spaces, $1 \leq p \leq \infty$, are complete with respect to the L^p-norm, this result will hold provided the sequence $\{s_n\}$ is Cauchy in L^2: $\lim_{n,m\to\infty} ||s_n - s_m||_2 = 0$. But if, say, $m \leq n$, the orthogonality relations (1.3) imply that

$$||s_n - s_m||_2^2 = (s_n, s_n) - 2(s_m, s_n) + (s_m, s_m)$$
$$= \sum_{n \geq |k| > m} |c_k|^2 = \sum_{n \geq |k| > m} |\hat{f}(k)|^2.$$

That this last sum tends to zero as m and n tend to ∞ follows from *Bessel's inequality* for functions f in $L^2(0, 1)$,

$$\sum_{k=-\infty}^{\infty} |\hat{f}(k)|^2 \leq ||f||_2^2 = \int_0^1 |f(x)|^2 \, dx, \tag{1.5}$$

which is an easy consequence of the orthogonality relations (1.3) and the definition of the Fourier coefficients $c_k = \hat{f}(k)$: since, for any g in $L^2(0, 1)$, $(g, g) \geq 0$, we have

$$0 \leq \left(\sum_{k=-n}^{n} c_k e_k - f, \sum_{k=-n}^{n} c_k e_k - f \right)$$
$$= \left(\sum_{k=-n}^{n} c_k e_k, \sum_{k=-n}^{n} c_k e_k \right)$$
$$- \left(\sum_{k=-n}^{n} c_k e_k, f \right) - \left(f, \sum_{k=-n}^{n} c_k e_k \right) + (f, f)$$
$$= \sum_{k=-n}^{n} |c_k|^2 - \sum_{k=-n}^{n} |c_k|^2 - \sum_{k=-n}^{n} |c_k|^2 + ||f||_2^2.$$

That is

$$\sum_{k=-n}^{n} |c_k|^2 \leq ||f||_2^2,$$

and (1.5) follows by letting $n \longrightarrow \infty$. These arguments give a flavor of the elegance and simplicity of the L^2-theory.

Another satisfactory solution of the problem of representation of functions by their Fourier series is to consider, instead of convergence, some methods of summability of Fourier series at individual points. The two best known types of summability (and the only ones we shall consider) are *Cesàro* and *Abel summability.* The former (often also referred to as the *method of summability by the first arithmetic means* or, simply, as *(C, 1) summability*) is defined in the following way: suppose we are given a numerical series $u_0 + u_1 + u_2 + \cdots$ with partial sums $s_0, s_1, s_2, \cdots$. We then form the *(C, 1) means* (or *first arithmetic means*)

$$\sigma_n = \frac{s_0 + s_1 + \cdots + s_n}{n+1} = \sum_{\nu=0}^{n} \left(1 - \frac{\nu}{n+1}\right) u_\nu$$

and say that the series is $(C, 1)$ summable to l if $\lim_{n\to\infty} \sigma_n = l$. The *Abel means* of the series are defined for each r, $0 \le r < 1$, by setting

$$A(r) = u_0 + u_1 r + u_2 r^2 + \cdots = \sum_{k=0}^{\infty} u_k r^k$$

and we say that the series is Abel summable to l if $\lim_{r\to 1-} A(r) = l$. It is not hard to show that if $u_0 + u_1 + u_2 + \cdots$ is convergent to the sum l then it must be both $(C, 1)$ and Abel summable to l. On the other hand, there are many series that are summable but not convergent. An illustrative example is the series $1 - 1 + 1 - 1 + \cdots = \sum_{k=0}^{\infty} (-1)^k$, whose $(C, 1)$ and Abel means are easily seen to converge to $\frac{1}{2}$. It can also be shown that $(C, 1)$ summability implies Abel summability. Thus, many results involving Abel summability follow from corresponding theorems that deal with Cesàro summability. Nevertheless, an independent study of the former is of interest, particularly when we consider Fourier series of functions f. This is true, not only because such series may be Abel summable under weaker conditions on f than are necessary to guarantee their $(C, 1)$ summability, but also because Abel summability has special properties, related to the theory of harmonic and analytic functions, that are not enjoyed by Cesàro summability.

We now describe, briefly, how these concepts apply to the study of Fourier series. The two most important results in connection with the problem of representing functions by their Fourier series are the following:

(1.6) *If f is periodic and integrable on* $(0, 1)$ *then the* $(C, 1)$ *means and the Abel means of the Fourier series of f converge to*

$$\tfrac{1}{2}\{f(x_0 + 0) + f(x_0 - 0)\}$$

at every point x_0 where the limits $f(x_0 \pm 0)$ exist. In particular, they converge at every point of continuity of f.

(1.7) *If f is periodic and integrable on* $(0, 1)$ *then the* $(C, 1)$ *means and the Abel means of the Fourier series of f converge to $f(x)$ for almost every x in* $(0, 1)$.†

We can obtain more insight into these results by examining more closely the first arithmetic means and the Abel means of the Fourier series of a function f. We first obtain an expression for the partial sums (1.2):

$$\begin{aligned} s_n(x) &= \sum_{k=-n}^{n} \hat{f}(k) e^{2\pi i k x} \\ &= \sum_{k=-n}^{n} \left(\int_0^1 f(t) e^{-2\pi i k t}\, dt \right) e^{2\pi i k x} \\ &= \int_0^1 \left(\sum_{k=-n}^{n} e^{2\pi i k (x-t)} \right) f(t)\, dt. \end{aligned}$$

By multiplying $D_n(\theta) = \sum_{k=-n}^{n} e^{2\pi i k \theta}$ by $2 \sin \pi\theta = i(e^{-i\pi\theta} - e^{i\pi\theta})$ all but the first and last term of the resulting sum cancel and we obtain

$$2\, D_n(\theta) \sin \pi\theta = i(e^{-(2n+1)\pi i \theta} - e^{(2n+1)\pi i \theta}) = 2 \sin (2n+1)\pi\theta;$$

† When restricted to Cesàro summability (1.6) is known as the theorem of Fejér and (1.7) as the theorem of Lebesgue. That these results then hold for Abel summability follows from the fact, mentioned above, that $(C, 1)$ summability implies Abel summability.

that is,

$$D_n(\theta) = \frac{\sin (2n+1)\pi\theta}{\sin \pi\theta}. \tag{1.8}$$

Hence,

$$\begin{aligned} s_n(x) &= \int_0^1 f(t)D_n(x-t)\,dt \\ &= \int_0^1 f(t)\frac{\sin (2n+1)\pi(x-t)}{\sin \pi(x-t)}\,dt. \end{aligned} \tag{1.9}$$

The expression (1.8) is called the *Dirichlet kernel.* We can now express the $(C, 1)$ means in terms of it:

$$\begin{aligned} \sigma_n(x) &= \frac{s_0(x) + s_1(x) + \cdots + s_n(x)}{n+1} \\ &= \frac{1}{n+1}\int_0^1 f(t)\left(\sum_{k=0}^{n} D_k(x-t)\right)dt. \end{aligned}$$

By multiplying the numerator and denominator of

$$K_n(\theta) = \frac{1}{n+1}\sum_{k=0}^{n} D_k(\theta) = \frac{1}{n+1}\sum_{k=0}^{n}\frac{\sin (2k+1)\pi\theta}{\sin \pi\theta}$$

by $\sin \pi\theta$ and replacing the products of sines in the numerator by differences of cosines we obtain

$$\begin{aligned} K_n(\theta) &= \frac{1}{n+1}\sum_{k=0}^{n}\frac{\cos 2k\pi\theta - \cos 2(k+1)\pi\theta}{2\sin^2 \pi\theta} \\ &= \frac{1}{n+1}\,\frac{1-\cos 2(n+1)\pi\theta}{2\sin^2 \pi\theta} \\ &= \frac{1}{n+1}\left[\frac{\sin (n+1)\pi\theta}{\sin \pi\theta}\right]^2. \end{aligned} \tag{1.10}$$

Consequently,

$$\begin{aligned} \sigma_n(x) &= \int_0^1 f(t)K_n(x-t)\,dt \\ &= \frac{1}{n+1}\int_0^1 f(t)\left\{\frac{\sin (n+1)\pi(x-t)}{\sin \pi(x-t)}\right\}^2 dt. \end{aligned} \tag{1.11}$$

$K_n(\theta)$ is called the *Fejér kernel.*

The result (1.6) follows easily from three basic properties of this kernel:

(A) $\int_0^1 K_n(\theta)\,d\theta = 1;$

(B) $K_n(\theta) \geq 0;$

(C) *for each* $\delta > 0$, $\max_{\delta \leq \theta \leq 1-\delta} K_n(\theta) \longrightarrow 0$ *as* $n \longrightarrow \infty$.

Property (B) is obvious. Property (A) is a consequence of the corresponding property for the Dirichlet kernel (which is immediate:

$$\int_0^1 D_n(\theta)\,d\theta = \sum_{k=-n}^{n} \int_0^1 e^{2\pi i k\theta}\,d\theta = 1)$$

and the representation $K_n(\theta) = \sum_{k=0}^{n} D_k(\theta)/(n+1)$. Finally, (C) follows from the inequality (see (1.10))

$$\max_{\delta \leq \theta \leq 1-\delta} K_n(\theta) \leq \frac{1}{n+1} \sin^{-2} \pi\delta.$$

Now, to obtain (1.6) we argue as follows: suppose x_0 is a point at which the limits $f(x_0 \pm 0)$ exist and let $a = \frac{1}{2}\{f(x_0 + 0) + f(x_0 - 0)\}$. Then, using the periodicity of the functions involved, the change of variables $t = x - s$, and property (A),

$$\begin{aligned}
\sigma_n(x_0) - a &= \int_{-1/2}^{1/2} f(s) K_n(x_0 - s)\,ds - a \cdot 1 \\
&= \int_{-1/2}^{1/2} f(x_0 - t) K_n(t)\,dt - a \int_{-1/2}^{1/2} K_n(t)\,dt \\
&= 2 \int_0^{\delta} \left\{ \frac{f(x_0 - t) + f(x_0 + t)}{2} - a \right\} K_n(t)\,dt \\
&\quad + \int_{\delta \leq |t| \leq 1/2} \{f(x_0 - t) - a\} K_n(t)\,dt.
\end{aligned}$$

Hence, if $\delta > 0$ is so chosen that $|f(x_0 - t) + f(x_0 + t) - 2a| \leq \epsilon$ if $|t| \leq \delta$, we have, by (B) and (A),

$$
\begin{aligned}
|\sigma_n(x_0) - a| &\leq \epsilon \int_0^\delta K_n(t)\,dt \\
&\quad + \left\{\max_{\delta \leq |t| \leq 1/2} K_n(t)\right\} \int_{\delta \leq |t| \leq 1/2} |f(x_0 - t) - a|\,dt \\
&\leq \epsilon \int_{-1/2}^{1/2} K_n(t)\,dt \\
&\quad + \left\{\max_{\delta \leq |t| \leq 1/2} K_n(t)\right\} \int_{-1/2}^{1/2} |f(x_0 - t) - a|\,dt \\
&= \epsilon \cdot 1 + \left\{\max_{\delta \leq |t| \leq 1/2} K_n(t)\right\} \int_{-1/2}^{1/2} |f(x_0 - t) - a|\,dt;
\end{aligned}
$$

but, by (C), the last term tends to 0 as $n \longrightarrow \infty$. Since $\epsilon > 0$ is arbitrary we can conclude that $\lim_{n\to\infty} |\sigma_n(x_0) - a| = 0$ and (1.6) is proved.

The theorem of Lebesgue, result (1.7) for the $(C, 1)$ means, is somewhat deeper and we postpone its proof until later. Since $(C, 1)$ summability implies Abel summability, as remarked above, both the results (1.6) and (1.7) follow once we establish them for Cesàro means. We commented before, however, that an independent study of Abel summability is of interest since it has special properties not enjoyed by $(C, 1)$ summability. This is easily made clear by examining the Abel means of Fourier series more closely; we do this by showing how the study of Fourier series is intimately connected with analytic function theory.

That this connection should exist is not surprising once we make the observation, *when f is real-valued*, that the series (1.1) is the real part of the power series

$$
\hat{f}(0) + \sum_{k=1}^{\infty} 2\hat{f}(k)z^k \tag{1.12}
$$

restricted to the unit circle $z = e^{2\pi i x}$. We note that this series defines an analytic function in the interior of the unit circle since the coefficients $\hat{f}(k)$ are uniformly bounded (in fact,

$$
|\hat{f}(k)| \leq \int_0^1 |f(t)|\,dt = ||f||_1).
$$

Thus, the real part of (1.12) is a harmonic function when $r =$

$|z| < 1$. But this real part is nothing more than the Abel mean of the Fourier series (1.1):

$$A(r, x) = A_f(r, x) = \hat{f}(0) + \sum_{k=1}^{\infty} r^k \hat{f}(k) e^{2\pi ikx} + \hat{f}(-k) e^{-2\pi ikx}$$

$$= \sum_{k=-\infty}^{\infty} r^{|k|} \hat{f}(k) e^{2\pi ikx}.$$

The imaginary part of (1.12), when $z = e^{2\pi ikx}$, is (formally),

$$(1.13) \qquad -i \sum_{k=-\infty}^{\infty} (\operatorname{sgn} k) \hat{f}(k) e^{2\pi ikx},$$

where, for any nonzero complex number z, $\operatorname{sgn} z = z/|z|$ and $\operatorname{sgn} 0 = 0$. This series is called the series *conjugate* to the Fourier series (1.1). Though it is not, in general, a Fourier series, this conjugate series is closely connected (see Sec. 4) to a (not necessarily integrable) function, the *conjugate function*, $\tilde{f}$.

As in the case of the $(C, 1)$ means, the Abel means $A(r, x)$ have an integral representation; that is, a representation similar to (1.11). We have, for $0 \leq r < 1$,

$$A(r, x) = \sum_{k=-\infty}^{\infty} r^{|k|} \hat{f}(k) e^{2\pi ikx}$$

$$= \sum_{k=-\infty}^{\infty} r^{|k|} \left(\int_0^1 f(t) e^{-2\pi ikt} \, dt \right) e^{2\pi ikx}$$

$$= \int_0^1 \left(\sum_{k=-\infty}^{\infty} r^{|k|} e^{2\pi ik(x-t)} \right) f(t) \, dt,$$

the change in the order of integration and summation being justifiable by the uniform convergence of the series

$$P(r, \theta) = \sum_{k=-\infty}^{\infty} r^{|k|} e^{2\pi ik\theta}$$

for $0 \leq r < 1$. But, setting $z = re^{2\pi i\theta}$, $P(r, \theta)$ is simply the real part of

$$1 + \sum_{k=1}^{\infty} 2r^k e^{2\pi ik\theta} = 1 + 2 \sum_{k=1}^{\infty} z^k = \frac{1+z}{1-z}.$$

Consequently,

$$(1.14) \qquad P(r, \theta) = \frac{1 - r^2}{1 - 2r \cos 2\pi\theta + r^2}$$

and we obtain the desired integral representation for the Abel means

$$(1.15) \quad A(r, x) = \int_0^1 P(r, x - t) f(t)\, dt$$

$$= \int_0^1 \frac{1 - r^2}{1 - 2r \cos 2\pi(x - t) + r^2} f(t)\, dt.$$

$P(r, \theta)$ is called the *Poisson kernel* and the integral (1.15) is called the *Poisson integral* of f. The reader can easily verify that this kernel satisfies the three properties, completely analogous to those of the Fejér kernel:

(A′) $\int_0^1 P(r, \theta)\, d\theta = 1$;

(B′) $P(r, \theta) \geq 0$;

(C′) *for each* $\delta > 0$, $\max_{\delta \leq \theta \leq 1 - \delta} P(r, \theta) \longrightarrow 0$ *as* $r \longrightarrow 1$.

From this we see that to the proof of (1.6) given above in the case of the Cesàro means there corresponds a practically identical proof of this result for the Abel means.

Let us observe that the imaginary part of $\dfrac{1 + z}{1 - z}$ has the form

$$Q(r, \theta) = \frac{2r \sin 2\pi\theta}{1 - 2r \cos 2\pi\theta + r^2}$$

and one readily obtains the Abel mean of the conjugate Fourier series (1.13) by the integral

$$(1.16) \quad \tilde{A}(r, x) = \int_0^1 Q(r, x - t) f(t)\, dt$$

$$= \int_0^1 \frac{2r \sin 2\pi(x - t)}{1 - 2r \cos 2\pi(x - t) + r^2} f(t)\, dt,$$

This integral is called the *conjugate Poisson integral* of f and $Q(r, \theta)$ is known as the *conjugate Poisson kernel*.

In this discussion we assumed that f was real-valued. It is clear.

however, that the Poisson integral formula (1.15) for the Abel means of the Fourier series of f holds in case f is complex-valued as well. To see this one need only apply it to the real and imaginary parts of f.

Before passing to other aspects of harmonic analysis let us examine more closely the integrals (1.9), (1.11), (1.15), and (1.16) that gave us the partial sums, the $(C, 1)$ means, the Abel means of the Fourier series of an integrable periodic function f, and the Abel means of the conjugate Fourier series of f. All these integrals have the form

$$(g * f)(x) = \int_0^1 g(x - t)f(t)\,dt, \tag{1.17}$$

where g is a periodic integrable function. In fact, in all these cases g is much better than merely integrable; for example, it is a bounded function and, consequently, it is obvious that, for each x, the integrand in (1.17) is integrable and $(g * f)(x)$ is well defined. We shall see, however, that the latter is well defined for almost all x when g is integrable. We therefore obtain a function, $g * f$ (defined almost everywhere), by forming the integral (1.17) whenever g and f belong to $L^1(0, 1)$ and are periodic. This operation, that assigns to each such pair (g, f) the function $g * f$, is called *convolution* and plays an important role in the theory of Fourier series. The most important elementary properties of convolution are the following:

(i) *If f and g are periodic and in $L^1(0, 1)$ so is $f * g$ and*

$$\begin{aligned} \|f * g\|_1 &= \int_0^1 |(f * g)(x)|\,dx \\ &\le \left(\int_0^1 |f(t)|\,dt\right)\left(\int_0^1 |g(t)|\,dt\right) \\ &= \|f\|_1\,\|g\|_1; \end{aligned}$$

(ii) $f * g = g * f$;

(iii) $(f * g) * h = f * (g * h)$ *whenever f, g, and h are periodic and in $L^1(0, 1)$;*

(iv) *For f, g, and h as in* (iii) *and any two complex numbers a and b*

$$f * (ag + bh) = a(f * g) + b(f * h).$$

That $(f * g)(x)$ is well defined for almost all x, as well as property (i), is an easy consequence of Fubini's theorem: since

$$|(f * g)(x)| \leq \int_0^1 |f(x-t)|\,|g(t)|\,dt$$

we have, using the periodicity of f,

$$\begin{aligned}\int_0^1 |(f * g)(x)|\,dx &\leq \int_0^1 \left(\int_0^1 |f(x-t)|\,|g(t)|\,dt\right) dx \\ &= \int_0^1 |g(t)| \left(\int_0^1 |f(x-t)|\,dx\right) dt \\ &= \int_0^1 |g(t)|\,\|f\|_1\,dt = \|f\|_1\,\|g\|_1.\dagger\end{aligned}$$

The remaining three properties follow from simple transformations of integrals and we omit their proofs.

The relation between Fourier transformation and convolution is very simple and elegant:

Suppose f and g are periodic and in $L^1(0, 1)$, then for all integers k

$$(f * g)^\wedge(k) = \hat{f}(k)\hat{g}(k).\ddagger \tag{1.18}$$

In order to see this we use Fubini's theorem and the periodicity of the functions involved:

$$\begin{aligned}(f * g)^\wedge(k) &= \int_0^1 \left(\int_0^1 f(x-t)g(t)\,dt\right) e^{-2\pi ikx}\,dx \\ &= \int_0^1 g(t)e^{-2\pi ikt} \left(\int_0^1 f(x-t)e^{-2\pi ik(x-t)}\,dx\right) dt \\ &= \int_0^1 g(t)e^{-2\pi ikt}\hat{f}(k)\,dt \\ &= \hat{f}(k)\hat{g}(k).\end{aligned}$$

---

† We are using the following version of Fubini's theorem: If $h \geq 0$ is a measurable function in the square $\{0 \leq x \leq 1,\ 0 \leq t \leq 1\}$ and the iterated integral $\int_0^1 (\int_0^1 h(x, t)\,dt)\,dx$ is finite, then h is integrable, $\int_0^1 h(x, t)\,dt$ is finite for almost every x and $\int_0^1 (\int_0^1 h(x, t)\,dt)\,dx = \int_0^1 (\int_0^1 h(x, t)\,dx)\,dt$. We have tacitly assumed, when $h(x, t) = |f(x-t)g(t)|$, that h is measurable. We leave the proof of this fact to the reader.

‡ In general, we shall let $(\quad)^\wedge$ denote the Fourier transform of the expression in the parentheses.

It is this result, that the Fourier transform of the convolution of two functions is, simply, the product of their Fourier transforms, that makes convolution play such an important role in the study of Fourier series. This, as we shall see, becomes clear very early in the development of harmonic analysis.

2. HARMONIC ANALYSIS ON THE INTEGERS AND ON THE REAL LINE

Up to this point we have considered only functions that were periodic of period 1. It is often useful to think of such functions as defined on the *additive group of real numbers modulo* 1† or on the perimeter of the unit circle $\{z \text{ complex};\ z = e^{2\pi i\theta}\}$ of the complex plane. Consequently, the theory of Fourier series is often referred to as the harmonic analysis associated with this circle, or the reals modulo 1. Toward the end of this monograph we shall describe how harmonic analysis can be associated to a wide variety of domains. In this section we shall consider two of these, the integers and the entire real line. The harmonic analysis corresponding to these domains is intimately connected with the theory of Fourier series.

In the case of the real line we obtain the theory of Fourier integrals, a topic that is as important and as well known as the theory of Fourier series. The harmonic analysis associated with functions defined on the integers, however, is not generally studied per se. The main reason is that its elementary aspects (but by no means its deeper ones) consist of results that are essentially on the surface. But precisely this property, this elementary nature of the subject, makes its study very worthwhile for the nonspecialist as it provides a great deal of motivation for the theories of Fourier series and integrals. Furthermore, some remarks about

---

† If we say that two real numbers are equivalent when their difference is an integer we obtain a partition of the reals into equivalence classes. Let $[x]$, $[y]$, $[z]$, $\cdots$ denote the equivalence classes containing the real numbers x, y, z, $\cdots$. Then the additive group of real numbers modulo 1 consists of these equivalence classes together with the operation defined by $[x] + [y] = [x + y]$.

this topic are necessary in order to understand better the general picture of our subject. Accordingly, we shall not consider this part of harmonic analysis in any detail, but will treat it briefly and use it to motivate the introduction of the inverse Fourier transform, which will enable us to consider the problem of representation of functions by their Fourier series from a more general point of view. Also, we shall use it to motivate our introductory remarks concerning Fourier integrals. We strongly urge the reader, however, to find the analogs, for functions defined on the integers, of results we shall present in the theories of Fourier series and integrals.

In the last section we started out with a periodic function belonging to $L^1(0, 1)$ and obtained, by means of a certain integral, a function defined on the integers, the Fourier transform. We then asked if it were possible to obtain the original function from the latter by means of a certain series, the Fourier series (1.1). This indicates a duality between the interval $(0, 1)$ and the integers and it is not unreasonable to expect that, by considering originally a function defined on the integers, we can introduce, in analogy to the Fourier transform, a periodic function by means of an appropriate series. Furthermore, we should be able to recapture the original function from this periodic function and a suitable integral. It is only natural, in view of these remarks, to hope that this can be done by interchanging the roles played by the interval $(0, 1)$ and the integers. More explicitly, let us examine the result of systematically replacing, in the definitions made at the beginning of the last section,

$$\sum_{k=-\infty}^{\infty} \text{ by } \int_0^1, \qquad \int_0^1 \text{ by } \sum_{k=-\infty}^{\infty},$$

the continuous variable $x \in (0, 1)$ by the integral variable k and k by x.

Let us consider, then, the integers as a measure space in which each point has measure 1 and an integrable function, f, defined on this measure space; that is, f satisfies

$$\sum_{k=-\infty}^{\infty} |f(k)| < \infty. \tag{2.1}$$

The space of such integrable functions is usually denoted by l^1. For f in l^1 we introduce the periodic function $\hat{f}$ whose value at x is

$$\hat{f}(x) = \sum_{j=-\infty}^{\infty} f(j)e^{-2\pi ijx}. \tag{2.2}$$

We shall call $\hat{f}$ the *Fourier transform* of f in this case as well. Because of the convergence (2.1) the series (2.2) converges uniformly; consequently, $\hat{f}$ is a continuous function. Corresponding to the Fourier series (1.1) we have the integral

$$\int_0^1 \hat{f}(x)e^{2\pi ikx}\,dx. \tag{2.3}$$

But the uniform convergence of (2.2), allowing us to integrate term-by-term, and the orthogonality relations (1.3) immediately imply that

$$\int_0^1 \hat{f}(x)e^{2\pi ikx}\,dx = f(k). \tag{2.4}$$

We therefore see that in the present case we do not encounter any of the difficulties described in the first section when we try to express the original function in terms of its Fourier transform. This illustrates the simplicity of the elementary aspects of the harmonic analysis associated with the integers.

In particular, we see that the mapping that assigns to $f \in l^1$ its Fourier transform is one-to-one and, thus, it has an inverse. This inverse, in view of (2.4), has an obvious extension to all of $L^1(0, 1)$; namely, the operator, called the *inverse Fourier transform,* that takes a function g in $L^1(0, 1)$ into the function $\check{g}$ whose value at $k = 0, \pm 1, \pm 2, \cdots$ is

$$\check{g}(k) = \int_0^1 g(x)e^{2\pi ikx}\,dx. \tag{2.5}$$

We can, therefore, rewrite (2.4) in the following way:

$$(\hat{f})^{\vee} = f, \tag{2.6}$$

whenever f belongs to l^1.

These considerations lead us to a useful and more general restatement of the problem we studied in the last section concerning the representation of functions by their Fourier series. Suppose we again interchange the roles played by the interval (0, 1) and the integers; it is then natural to try to define the inverse Fourier transform of a function, g, whose domain is the integers, by the expression

$$\check{g}(x) = \sum_{k=-\infty}^{\infty} g(k)e^{2\pi ikx}. \tag{2.7}$$

When g is in l^1 the series (2.7) is convergent and we obtain a well-defined mapping, $g \longrightarrow \check{g}$, from l^1 into the class of continuous periodic functions. This mapping, however, is insufficient for our purposes. For example, in view of (2.6), we should expect that whenever g is the Fourier transform of an integrable function f it then follows that $\check{g} = (\hat{f})^\vee = f$. But, because of Kolmogoroff's example of an integrable function with an everywhere divergent Fourier series, this equality cannot be valid if we let $\check{g}$ be defined as the function which, at each x (or at almost every x), satisfies (2.7) in the usual sense (that is, the sequence of partial sums $s_n(x) = \sum_{k=-n}^{n} g(k)e^{2\pi ikx}$ converges). On the other hand, using (1.7), we do obtain an almost everywhere defined $\check{g}(x) = (\hat{f})^\vee(x) = f(x)$ if we interpret (2.7) to mean that the $(C, 1)$ or Abel means of the series on the right converge to $\check{g}(x)$. Similarly, if g is the Fourier transform of an f belonging to $L^2(0, 1)$, (2.7) gives us a well-defined function $\check{g} = f$ if we interpret the series on the right to be convergent in the L^2-norm. In each of these cases we obtain a mapping which is an inverse to the Fourier transform mapping when the latter is restricted to some important domain of functions (such as $L^2(0, 1)$ or $L^1(0, 1)$). Thus, a general formulation of the problem we discussed in the last section is the following: given a class C of periodic functions for which the Fourier transform is defined, does there exist a mapping, $g \longrightarrow \check{g}$, defined on a class of functions, whose common domain is the integers, such that $(\hat{f})^\vee$ is defined for all f in C and $(\hat{f})^\vee = f$? We shall refer to the problem stated in this form as the *Fourier inversion problem.*

Let us now turn to the harmonic analysis related to the real line, the theory of Fourier integrals. Suppose that f is integrable over $(-\infty, \infty)$, then its *Fourier transform* is defined for all real x by

$$\hat{f}(x) = \int_{-\infty}^{\infty} f(t)e^{-2\pi ixt}\,dt.$$

The integral on the right is usually also called the Fourier integral. Since the Fourier transform of a function defined on the entire real line is again such a function it should not surprise the reader, in view of our discussion in this section, that a good heuristic approach for obtaining the basic notions and results in the theory of Fourier integrals is to let the real line play the roles that the interval $(0, 1)$ and the integers played in the theory of Fourier series. Let us examine the Fourier inversion problem from this point of view.

Motivated by the expressions (2.5), (2.6), and (2.7) we would expect that the inverse Fourier transform of a function, g, defined on the real line should be given by the formula

$$\check{g}(x) = \int_{-\infty}^{\infty} g(t)e^{2\pi ixt}\,dt, \tag{2.8}$$

and that for each f in $L^1(-\infty, \infty)$ we would then have $(\hat{f})^\vee = f$. Just as in the case of Fourier series, however, we immediately encounter the problem of giving relation (2.8) a suitable interpretation. Though $\hat{f}$ has several nice properties (the reader can easily check that it is uniformly continuous and bounded; in fact,

$$\|\hat{f}\|_\infty = \sup_{-\infty<x<\infty} |\hat{f}(x)| \leq \int_{-\infty}^{\infty} |f(t)|\,dt = \|f\|_1 \tag{2.9}$$

whenever f is in $L^1(-\infty, \infty)$) it is not always true that it is integrable. For example, if f is the characteristic function, $X_{(a,b)}$, of the finite interval (a, b) then

$$\hat{X}_{(a,b)}(x) = \int_a^b e^{-2\pi ixt}\,dt = \frac{e^{-2\pi ixa} - e^{-2\pi ixb}}{2\pi ix}, \tag{2.10}$$

when $x \neq 0$ and $\hat{X}_{(a,b)}(0) = b - a$. Here, as in the last section, we obtain a satisfactory solution of the Fourier inversion problem if we consider $(C, 1)$ and Abel summability. We see this easily if we let ourselves be guided by the above mentioned heuristic principle of substituting the real line for the interval $(0, 1)$ and the integers. Let us first examine briefly the case of Cesàro summability.

If u is integrable in the intervals $[-R, R]$, for all $R > 0$, the

Cesàro means, or *(C, 1) means*, of $\int_{-\infty}^{\infty} u(t)\,dt$ are defined by the integrals

$$\sigma_R = \int_{-R}^{R} \left(1 - \frac{|t|}{R}\right) u(t)\,dt.$$

We say that $\int_{-\infty}^{\infty} u(t)\,dt$ is $(C, 1)$ summable to l if $\lim_{R\to\infty} \sigma_R = l$. It is easy to see that if $u \in L^1(-\infty, \infty)$ and its integral is l then $\sigma_R \longrightarrow l$ as $R \longrightarrow \infty$.

Let us now consider the Cesàro means of the integral in (2.8) when g is the Fourier transform of an integrable function f. We have

$$\begin{aligned}\sigma_R(x) &= \int_{-R}^{R} \left(1 - \frac{|t|}{R}\right) e^{2\pi i x t}\hat{f}(t)\,dt \\ &= \int_{-R}^{R} \left(1 - \frac{|t|}{R}\right) e^{2\pi i x t} \left\{\int_{-\infty}^{\infty} f(y) e^{-2\pi i t y}\,dy\right\} dt \\ &= \int_{-\infty}^{\infty} f(y) \left\{\int_{-R}^{R} \left(1 - \frac{|t|}{R}\right) e^{2\pi i t(x-y)}\,dt\right\} dy.\end{aligned}$$

It is not hard to obtain a simple expression for the inner integral. Using the fact that the sine function is odd and integrating by parts we obtain:

$$\begin{aligned}&K_R(\theta) \\ &= \int_{-R}^{R} \left(1 - \frac{|t|}{R}\right) e^{2\pi i \theta t}\,dt = 2\int_{0}^{R} \left(1 - \frac{t}{R}\right) \cos(2\pi\theta t)\,dt \\ &= 2\int_{0}^{R} \frac{1}{R} \frac{\sin(2\pi\theta t)}{2\pi\theta}\,dt = \frac{1}{2\pi^2 R} \cdot \frac{1 - \cos(2\pi R\theta)}{\theta^2}\end{aligned} \tag{2.11}$$

Consequently,

$$\begin{aligned}\sigma_R(x) &= \int_{-\infty}^{\infty} f(y) K_R(x - y)\,dy \\ &= \frac{1}{2\pi^2 R} \int_{-\infty}^{\infty} f(y) \frac{1 - \cos(2\pi R(x - y))}{(x - y)^2}\,dy.\end{aligned} \tag{2.12}$$

$K_R(\theta)$ is called the *Fejér kernel* and it satisfies the following three basic properties:

(a) $\int_{-\infty}^{\infty} K_R(\theta)\, d\theta = 1;$

(b) $K_R(\theta) \geq 0;$

(c) *for each* $\delta > 0$, $\int_{|\theta| \geq \delta} K_R(\theta)\, d\theta \longrightarrow 0$ *as* $R \longrightarrow \infty$.

The second property is obvious. The first property follows easily from the well-known result: $\lim_{N\to\infty} \int_0^N (\sin t/t)\, dt = \pi/2$. To see this we change variables and integrate by parts:

$$\int_{-\infty}^{\infty} K_R(\theta)\, d\theta = 2\int_{-\infty}^{\infty} \frac{1 - \cos 2\pi R\theta}{R(2\pi\theta)^2}\, d\theta = \frac{1}{\pi}\int_{-\infty}^{\infty} \frac{1-\cos s}{s^2}\, ds$$

$$= \frac{2}{\pi}\lim_{N\to\infty} \int_0^N \frac{1-\cos s}{s^2}\, ds = \frac{2}{\pi}\lim_{N\to\infty}\int_0^N \frac{\sin t}{t}\, dt = 1.$$

In order to prove (c) let us first observe that $K_R(\theta) \leq 1/R\theta^2$ (thus,

$$\max_{|\theta|\geq\delta} K_R(\theta) \leq \frac{1}{R}\max_{|\theta|\geq\delta}\frac{1}{\theta^2} \longrightarrow 0 \tag{2.13}$$

as $R \longrightarrow \infty$). Consequently,

$$\int_{|\theta|\geq\delta} K_R(\theta)\, d\theta \leq \frac{1}{R}\int_{|\theta|\geq\delta}\frac{d\theta}{\theta^2} = \frac{2}{R\delta} \longrightarrow 0 \qquad \text{as } R \longrightarrow \infty.$$

If we replace (c) by (2.13) we have three properties that are completely analogous to the properties (A), (B), and (C) of the Fejér kernel obtained in the periodic case. Precisely the same argument that is used in establishing the theorem of Fejér (see (1.6)) will then give us the corresponding result for Fourier integrals. We introduce property (c), however, to show how $(C, 1)$ summability can be used in yet another way in order to obtain a solution of the Fourier inversion problem. More precisely, we shall prove the following result:

(2.14) *If f is integrable then the $(C, 1)$ means*

$$\sigma_R(x) = \int_{-R}^{R}\left(1 - \frac{|t|}{R}\right) e^{2\pi i x t}\hat{f}(t)\, dt = \int_{-\infty}^{\infty} f(t)K_R(x-t)\, dt$$

of the integral defining the inverse Fourier transform of $\hat{f}$ converge to f in the L^1 norm. That is,

$$\lim_{R\to\infty} ||f - \sigma_R||_1 = \lim_{R\to\infty} \int_{-\infty}^{\infty} |\sigma_R(x) - f(x)|\, dx = 0.$$

It is convenient, at this point, to introduce the L^p *modulus of continuity* of a function f in $L^p(-\infty, \infty)$:

$$\omega_p(\delta) = \max_{0\le t\le\delta} \left\{\int_{-\infty}^{\infty} |f(x+t) - f(x)|^p\, dx\right\}^{1/p}.$$

It is an elementary fact that $\omega_p(\delta) \longrightarrow 0$ as $\delta \longrightarrow 0$ when $1 \le p < \infty$. For, given $\delta > 0$, we can write $f = f_1 + f_2$, where f_1 is continuous, vanishes outside a finite interval, and $||f_2||_p < \epsilon/3$. Thus, by Minkowski's inequality,

$$\begin{aligned}\left\{\int_{-\infty}^{\infty} |f(x+t) - f(x)|^p\, dx\right\}^{1/p} &\\ \le \left\{\int_{-\infty}^{\infty} |f_1(x+t) - f_1(x)|^p\, dx\right\}^{1/p} &\\ + \left\{\int_{-\infty}^{\infty} |f_2(x+t)|^p\, dx\right\}^{1/p} + \left\{\int_{-\infty}^{\infty} |f_2(x)|^p\, dx\right\}^{1/p}&.\end{aligned}$$

Each of the last two terms is less than $\epsilon/3$; since f_1 is uniformly continuous and vanishes outside a finite interval, the last term is also smaller than $\epsilon/3$ provided t is close enough to 0. Thus, $\omega_p(\delta) < \epsilon/3 + \epsilon/3 + \epsilon/3 = \epsilon$ if δ is close enough to 0.

We now prove (2.14). Using the change of variables $t = x - s$ and property (a),

$$\begin{aligned}\sigma_R(x) - f(x) &= \int_{-\infty}^{\infty} f(x)K_R(x-s)\, ds - f(x)\cdot 1\\ &= \int_{-\infty}^{\infty} [f(x-t) - f(x)]K_R(t)\, dt.\end{aligned}$$

Thus,

$$\begin{aligned}\int_{-\infty}^{\infty} |\sigma_R(x) - f(x)|\, dx &= \int_{-\infty}^{\infty} \left|\int_{-\infty}^{\infty} [f(x-t) - f(x)]K_R(t)\, dt\right| dx\\ &\le \int_{-\infty}^{\infty} \left\{\int_{-\infty}^{\infty} |f(x-t) - f(x)|\, K_R(t)\, dt\right\} dx\\ &= \int_{-\infty}^{\infty} \left\{\int_{-\infty}^{\infty} |f(x-t) - f(x)|\, dx\right\} K_R(t)\, dt\\ &= \int_{|t|\le} \left\{\int_{-\infty}^{\infty} |f(x-t) - f(x)|\, dx\right\} K_R(t)\, dt\\ &\quad + \int_{|t|>\delta} \left\{\int_{-\infty}^{\infty} |f(x-t) - f(x)|\, dx\right\} K_R(t)\, dt.\end{aligned}$$

The first of these last two terms is clearly dominated by

$$\omega_1(\delta) \int_{-\infty}^{\infty} K_R(t)\, dt = \omega_1(\delta),$$

which tends to 0 as δ tends to 0. The second term is majorized by

$$\int_{|t|>\delta} \left\{ \int_{-\infty}^{\infty} |f(x-t)|\, dx + \int_{-\infty}^{\infty} |f(x)|\, dx \right\} K_R(t)\, dt$$
$$= 2||f||_1 \int_{|t|>\delta} K_R(t)\, dt.$$

Thus, given $\epsilon > 0$, let us first choose $\delta > 0$ so that $\omega_1(\delta) < \epsilon/2$; then, with this δ fixed, property (c) can be used to find $R_0 > 0$ so that

$$\int_{|t|>\delta} K_R(t)\, dt < \frac{\epsilon}{4||f||_1}$$

when $R \geq R_0$. This shows that

$$\int_{-\infty}^{\infty} |\sigma_R(x) - f(x)|\, dx < \frac{\epsilon}{2} + \frac{2||f||_1}{4||f||_1}\epsilon = \epsilon,$$

provided $R \geq R_0$, and (2.14) is proved.

We recall that in the case of Fourier series the Abel means behaved very much like the $(C, 1)$ means, yet an independent study of them was of interest, particularly when we examined the relation between the theory of Fourier series and the theory of harmonic and analytic functions of a complex variable. This is equally true for Fourier integrals; consequently it is worthwhile, at this point, to devote a few words to Abel summability and its relation to the Fourier inversion problem.

Guided by our heuristic principle of replacing the integers by the real line and sums by integrals, we would expect the Abel means of the integral $\int_{-\infty}^{\infty} u(t)\, dt$ to be defined by the expression $\int_{-\infty}^{\infty} r^{|t|}u(t)\, dt$ with $0 \leq r < 1$. For technical reasons, which will become apparent shortly, it is convenient to put $r = e^{-2\pi y}$, $0 < y < \infty$; thus the Abel means of $\int_{-\infty}^{\infty} u(t)\, dt$ have the form

$$A(y) = \int_{-\infty}^{\infty} e^{-2\pi y|t|}u(t)\, dt, \qquad y > 0,$$

and we say that our integral is Abel summable to l if $\lim_{y \to 0+} A(y) = l$.

Let us now examine the Abel means of the integral (2.8) when g is the Fourier transform of an integrable function f. We then have

$$\begin{aligned} f(x, y) &= \int_{-\infty}^{\infty} e^{-2\pi y|t|} e^{2\pi i x t} \hat{f}(t)\, dt \\ &= \int_{-\infty}^{\infty} e^{-2\pi y|t|} e^{2\pi i x t} \left\{ \int_{-\infty}^{\infty} f(s) e^{-2\pi i t s}\, ds \right\} dt \\ &= \int_{-\infty}^{\infty} f(s) \left\{ \int_{-\infty}^{\infty} e^{-2\pi y|t|} e^{2\pi i t(x-s)}\, dt \right\} ds. \end{aligned}$$

As in the case of the $(C, 1)$ means we can easily obtain a simple expression for the inner integral:

$$\begin{aligned} \int_{-\infty}^{\infty} e^{-2\pi y|t|} e^{2\pi i t x}\, dt &= \int_{0}^{\infty} e^{2\pi t(ix-y)}\, dt + \int_{-\infty}^{0} e^{2\pi t(ix+y)}\, dt \\ &= \frac{1}{2\pi(y - ix)} + \frac{1}{2\pi(y + ix)} = \frac{1}{\pi} \frac{y}{x^2 + y^2}. \end{aligned}$$

Hence,

$$f(x, y) = \frac{1}{\pi} \int_{-\infty}^{\infty} f(t) \frac{y}{(x - t)^2 + y^2}\, dt \tag{2.15}$$

$$= \int_{-\infty}^{\infty} f(t) P(x - t, y)\, dt$$

where

$$P(x, y) = \frac{1}{\pi} \frac{y}{x^2 + y^2}, \tag{2.16}$$

for $y > 0$ and $-\infty < x < \infty$. $P(x, y)$ is called the *Poisson kernel* and the integral (2.15) is called the *Poisson integral of f*.

It is clear that result (2.14) still holds if we replace the Cesàro means of the integral $\int_{-\infty}^{\infty} \hat{f}(t) e^{2\pi i x t}\, dt$ by the Abel means, provided we can show that the Poisson kernel satisfies

(a′) $\int_{-\infty}^{\infty} P(x, y)\, dx = 1$;

(b′) $P(x, y) \geq 0$;

(c′) *for each* $\delta > 0$ $\int_{|x| \geq \delta} P(x, y)\, dy \longrightarrow 0$ *as* $y \longrightarrow 0$.

But the last two of these properties are obvious. Property (a′) is also easy to establish: if we let $s = x/y$ then

$$\int_{-\infty}^{\infty} P(x, y)\, dx = \frac{1}{\pi}\int_{-\infty}^{\infty} \frac{ds}{1 + s^2}$$

$$= \lim_{N\to\infty} \frac{1}{\pi} [\tan^{-1} N - \tan^{-1}(-N)] = 1.$$

We note that the Poisson kernel is a harmonic function in the upper half-plane $\{z = x + iy; y > 0\}$. This can be seen either by computing its Laplacian directly or by observing that it is the real part of the analytic function

$$\frac{i}{\pi}\frac{1}{z} = \frac{1}{\pi}\frac{y}{x^2 + y^2} + i\frac{1}{\pi}\frac{x}{x^2 + y^2}.$$

The imaginary part,

$$Q(x, y) = \frac{1}{\pi}\frac{x}{x^2 + y^2},$$

is called the *conjugate Poisson kernel.*

Now suppose f belongs to $L^1(-\infty, \infty)$ and is real-valued. Let us form the integral

$$F(z) = F(x + iy) = \frac{i}{\pi}\int_{-\infty}^{\infty} f(t)\frac{1}{(x - t) + iy}\, dt, \tag{2.17}$$

where $y > 0$ and $-\infty < x < \infty$. It is easy to see that F is an analytic function in the upper half-plane† and that its real part is given by the Poisson integral of f. Thus the latter defines a harmonic function in the upper half-plane. The imaginary part,

$$\tilde{f}(x, y) = \frac{1}{\pi}\int_{-\infty}^{\infty} f(t)\frac{x - t}{(x - t)^2 + y^2}\, dt \tag{2.18}$$

$$= \int_{-\infty}^{\infty} f(t)Q(x - t, y)\, dt,$$

† Perhaps the easiest way of proving this fact is to use Morera's theorem: given a simple closed contour C in the upper half-plane we can easily check that

$$\int_C F(z)\, dz = \frac{i}{\pi}\int_{-\infty}^{\infty} f(t)\left\{\int_C \frac{dz}{z - t}\right\} dt = \frac{i}{\pi}\int_{-\infty}^{\infty} f(t)\cdot 0\, dt = 0.$$

This then implies the analyticity of F. We leave the details of this argument to the reader.

is called the *conjugate Poisson integral* of f. We see from these observations that there must be a strong connection between Poisson integrals and the theory of harmonic and analytic functions of a complex variable.

By now we have given a good deal of evidence that the theory of Fourier integrals is not only intimately connected with the theory of Fourier series but is very similar to it. This is indeed the case. We shall see in more detail in the next section, for example, how the L^2-theory of Fourier integrals is as elegant as its analog, the L^2-theory of Fourier series, which we described briefly in the first section. We shall not, however, always discuss a result concerning, say, Fourier series and then also describe the corresponding result in the theory of Fourier integrals. On the contrary, we shall often discuss an aspect of harmonic analysis in one of the two theories, but not in both. The reader should be aware that there exist parallel results in the other theory as well. For example, the operation of convolution of two functions in $L^1(-\infty, \infty)$ plays an equally important role on the real line. Its definition is the obvious one: if f and g are integrable on $(-\infty, \infty)$, then $f * g$ is defined by

$$(f * g)(x) = \int_{-\infty}^{\infty} f(t)g(x - t)\, dt.$$

We leave it to the reader to check that $(f * g)(x)$ is defined for almost all real x and that the properties (i), (ii), (iii), and (iv) announced in the first section hold in this case as well. Moreover, the argument that was used to establish the important relation (1.18) between convolution and Fourier transformation can be used, after some obvious changes, to show that the same relation holds in the case of Fourier integrals.

3. THE L^1 AND L^2 THEORIES

In the last two sections we introduced several concepts but have not studied any of them very deeply. In this section we shall examine in much greater detail the convergence and summability of Fourier series, the Fourier inversion problem, and the L^2-theory. Moreover, we shall describe some of the better-known theorems in the harmonic analysis associated with the real line and the circle.

Let us begin with a closer look at the Fourier coefficients of an integrable and periodic function f. We have seen that when f belongs to $L^2(0, 1)$ the Fourier coefficients $c_k = \hat{f}(k)$ satisfy Bessel's inequality (see (1.5)). Thus, in particular,

$$\sum_{k=-\infty}^{\infty} |c_k|^2 < \infty .$$

It follows, therefore, that $c_k \longrightarrow 0$ as $|k| \longrightarrow \infty$. But this result is true even when f is in $L^1(0, 1)$. For, suppose $\epsilon > 0$, we can then write $f = g + h$ where g is in $L^2(0, 1)$ and $||h||_1 < \epsilon/2$. Since $|\hat{h}(k)| \leq ||h||_1$ for $k = 0, \pm 1, \pm 2, \cdots$, we have $|\hat{f}(k)| \leq |\hat{g}(k)| + \epsilon/2$. Now, using the result just established for $L^2(0, 1)$, we can find an $N > 0$ such that $|\hat{g}(k)| < \epsilon/2$ if $|k| \geq N$. Thus, $|\hat{f}(k)| < \epsilon/2 + \epsilon/2 = \epsilon$ if $|k| \geq N$. We have proved the following result:

(3.1) THE RIEMANN-LEBESGUE THEOREM: *If f is an integrable and periodic function then* $\lim_{|k| \to \infty} \hat{f}(k) = 0$.

An immediate application of the Riemann-Lebesgue theorem is the following convergence test for Fourier series.

(3.2) DINI TEST: *If a periodic and integrable function, f, satisfies the condition*

$$\int_{-1/2}^{1/2} \left| \frac{f(x-t) - f(x)}{\tan \pi t} \right| dt < \infty \tag{3.3}$$

at a point x, then the partial sums $s_n(x) = \sum_{k=-n}^{n} \hat{f}(k) e^{2\pi i k x}$ *converge to $f(x)$ as $n \longrightarrow \infty$.*

To see this we let g be the function whose value at $t \in (-\frac{1}{2}, \frac{1}{2})$ is $[f(x-t) - f(x)]/\tan \pi t$, then the integrability of f and condition (3.3) imply that g is integrable. Using (1.9) and the fact that

$$\int_{-1/2}^{1/2} D_n(t)\, dt = 1$$

we have, for $n \geq 1$,

$$\begin{aligned}
s_n(x) &- f(x) \\
&= \int_{-1/2}^{1/2} f(x-t)\,\frac{\sin(2n+1)\pi t}{\sin \pi t}\,dt - f(x)\int_{-1/2}^{1/2} D_n(t)\,dt \\
&= \int_{-1/2}^{1/2} \{f(x-t) - f(x)\}\,\frac{\sin(2n+1)\pi t}{\sin \pi t}\,dt \\
&= \int_{-1/2}^{1/2} \{f(x-t) - f(x)\}\left\{\frac{e^{2\pi int} - e^{-2\pi int}}{2i\tan \pi t} + \frac{e^{2\pi int} + e^{-2\pi int}}{2}\right\}dt \\
&= \frac{\hat{g}(-n) - \hat{g}(n)}{2i} + \frac{e^{-2\pi inx}\hat{f}(-n) + e^{2\pi inx}\hat{f}(n)}{2}.
\end{aligned}$$

But it follows from the Riemann-Lebesgue theorem that this last expression tends to 0 as $n \longrightarrow \infty$. This proves (3.2).

The Dini test is probably the most useful of the various convergence criteria in the literature. One of its consequences is the fact that the *Fourier series of an integrable and periodic function converges to the value of the function at each point of differentiability.* We see this by first noting that, since $\lim_{t\to 0} \frac{\tan \pi t}{\pi t} = 1$, condition (3.3) is equivalent to the condition

$$\text{(3.3}'\text{)} \qquad \int_{-1/2}^{1/2} \left|\frac{f(x-t) - f(x)}{t}\right| dt < \infty.$$

But it is obvious that if f is differentiable at x then (3.3′) must hold.

Before passing to the topic of summability of Fourier series we state, without proof, what is probably the best-known convergence test in the theory of Fourier series:

(3.4) THE DIRICHLET-JORDAN TEST: *Suppose a periodic function f is of bounded variation over* (0, 1). *Then*

(a) *the partial sums $s_n(x)$ converge to $\frac{1}{2}\{f(x+0) + f(x-0)\}$ at each real number x. In particular, they converge to $f(x)$ at each point of continuity of f;*

(b) *if f is continuous on a closed interval then $s_n(x)$ converges uniformly on this interval.*

We now pass to a more detailed study of summability of Fourier series. Let us observe that in the proof of the theorem of Fejér (result (1.6) restricted to Cesàro summability) we really have shown that the convergence of $\sigma_n(x)$ to $f(x)$ is uniform in any interval where f is uniformly continuous. From this we easily obtain the following classical result:

(3.5) WEIERSTRASS APPROXIMATION THEOREM: *Suppose f is a continuous periodic function and $\epsilon > 0$. Then there exists a trigonometric polynomial T, that is, a finite linear combination of the exponentials $e^{2\pi inx}$, $n = 0, \pm 1, \pm 2, \cdots$, such that*

$$|f(x) - T(x)| < \epsilon \qquad \textit{for all } x.$$

For we may take $T(x) = \sigma_n(x)$ for n large enough since the Cesàro means converge to f uniformly in this case.

One important consequence of these considerations is that *the system $\{e^{2\pi inx}\}$ is complete;* that is, *if all the Fourier coefficients of an integrable periodic function f vanish then f must be 0 almost everywhere.* We first note that if f is continuous and $\hat{f}(k) = 0$ for all integers k then $\sigma_n(x) \equiv 0$ for all n. But, since $\sigma_n(x) \longrightarrow f(x)$ at all x, we must have $f(x) \equiv 0$. If f is merely integrable and periodic we form the indefinite integral $F(x) = \int_0^x f(t)\,dt$. The condition $\hat{f}(0) = 0$ implies that

$$F(x+1) - F(x) = \int_x^{x+1} f(t)\,dt = \int_0^1 f(t)\,dt = 0.$$

Thus, F is a continuous periodic function. We claim that the hypothesis $\hat{f}(k) = 0$ for $k = \pm 1, \pm 2, \pm 3, \cdots$ implies that $\hat{F}(k) = 0$ for $k = \pm 1, \pm 2, \pm 3, \cdots$. For, integrating by parts,

$$\hat{F}(k) = \int_0^1 F(t)e^{-2\pi ikt}\,dt = F(t)\frac{e^{-2\pi ikt}}{-2\pi ik}\Big]_0^1 + \frac{1}{2\pi ik}\int_0^1 f(t)e^{-2\pi ikt}\,dt$$

$$= 0 + \frac{\hat{f}(k)}{2\pi ik} = 0 + 0 = 0.$$

From this and the orthogonality relations (1.3) we then conclude that the continuous and periodic function G whose value at x is $F(x) - \hat{F}(0)$ must satisfy $\hat{G}(k) = 0$ for all integers k. But we have shown that this implies that $G(x) \equiv 0$. Since, by Lebesgue's theorem on the differentiation of the integral $F'(x) = f(x)$ for almost every x, it follows that $0 = G'(x) = F'(x) = f(x)$ almost everywhere. This proves the completeness of the system $\{e^{2\pi inx}\}$.

We now show how to obtain the theorem of Lebesgue (see result (1.7)) that asserts that the $(C, 1)$ means of the Fourier series of an integrable and periodic function converge almost everywhere to the values of the function. In order to do this we will have to introduce the *Lebesgue set* of such a function f. We have just used the well-known fact that $F'(x) = f(x)$ for almost every x when $F(x) = \int_0^x f(t)\,dt$. We can rewrite this fact in the following way:

$$\lim_{h\to 0} \frac{1}{h} \int_0^h \{f(x+t) - f(x)\}\,dt = 0$$

for almost every x. It turns out that a stronger result is true:

$$\lim_{h\to 0} \frac{1}{h} \int_0^h |f(x+t) - f(x)|\,dt = 0 \tag{3.6}$$

for almost every x. It is not hard to show this: For a fixed rational number r let E_r be the set of all x such that

$$\lim_{h\to 0} \frac{1}{h} \int_0^h |f(x+t) - r|\,dt = |f(x) - r|$$

fails to hold. Applying Lebesgue's theorem on the differentiation of the integral to $g(t) = |f(t) - r|$ we conclude that E_r has measure 0. Let $E = \bigcup E_r$, the union being taken over all rational numbers r. Then E also has measure 0. We claim that if x does not belong to E then (3.6) holds. For let $\epsilon > 0$. Choose a rational number r_0 such that $|f(x) - r_0| < \epsilon/2$. Then

$$\frac{1}{h} \int_0^h |f(x+t) - f(x)|\,dt \leq \frac{1}{h} \int_0^h |f(x+t) - r_0|\,dt + \frac{1}{h} \int_0^h |f(x) - r_0|\,dt.$$

But the first term of this sum is less than $\epsilon/2$ if h is close to 0 while

$$\frac{1}{h} \int_0^h |f(x) - r_0|\,dt < \frac{1}{h} \int_0^h \frac{\epsilon}{2}\,dt = \frac{\epsilon}{2}.$$

Thus

$$\frac{1}{h} \int_0^h |f(x+t) - f(x)|\,dt < \epsilon$$

if h is small.

The set of all x such that (3.6) holds is called the *Lebesgue set* of f. We shall show that the $(C, 1)$ means of the Fourier series of f converge to $f(x)$ whenever x is a member of the Lebesgue set.

We shall need the following two estimates on the Fejér kernel:

(a) $K_n(t) \leq n + 1$;

(b) $K_n(t) \leq \dfrac{A}{(n+1)t^2}$, $|t| \leq \dfrac{1}{2}$, where A is an absolute constant.

The first one follows from the obvious estimate on the Dirichlet kernel $|D_k(t)| \leq 2k + 1$: for

$$K_n(t) = \frac{1}{n+1} \sum_{k=0}^{n} D_k(t) \leq \frac{1}{n+1} \sum_{k=0}^{n} (2k+1)$$

$$= \frac{(n+1)^2}{n+1} = n + 1.$$

The second one is a consequence of formula (1.10) and the well-known fact $\lim\limits_{t \to 0} \dfrac{\sin t}{t} = 1$.

Now suppose x belongs to the Lebesgue set of f. As in the proof of (1.6), we have, using property (A) of the Fejér kernel,

$$\sigma_n(x) - f(x) = \int_{-1/2}^{1/2} \{f(x-t) - f(x)\} K_n(t)\, dt.$$

Thus, using the estimates (a) and (b),

$$\begin{aligned} |\sigma_n(x) - f(x)| &\leq \int_{-1/2}^{1/2} |f(x-t) - f(x)| K_n(t)\, dt \\ &\leq (n+1) \int_{|t| \leq 1/(n+1)} |f(x-t) - f(x)|\, dt \\ &\quad + \frac{A}{n+1} \int_{1/(n+1) \leq |t| \leq 1/2} \frac{|f(x-t) - f(x)|}{t^2}\, dt. \end{aligned}$$

Given $\epsilon > 0$ let $\delta > 0$ be such that $\dfrac{1}{h} \displaystyle\int_{|t| \leq h} |f(x-t) - f(x)|\, dt < \epsilon$ if $h \leq \delta$. Then the first term in the above sum is less than ϵ whenever $(n+1)^{-1} \leq \delta$. In order to estimate the second term we write the integral as the sum of the two integrals $\displaystyle\int_{(n+1)^{-1}}^{1/2}$ and $\displaystyle\int_{-1/2}^{-(n+1)^{-1}}$. We shall show that the first integral tends to 0 as $n \longrightarrow \infty$; a similar estimate will then show that the same is true for the second.

Let $G(t) = \int_0^t |f(x-s) - f(x)|\, ds$. Then, integrating by parts, we have

$$\frac{A}{(n+1)} \int_{1/(n+1)}^{1/2} \frac{|f(x-t) - f(x)|}{t^2}\, dt$$

$$\leq \frac{4A}{n+1} G(1/2) + \frac{2A}{n+1} \int_{1/(n+1)}^{\delta} \frac{G(t)}{t^3}\, dt + \frac{2A}{n+1} \int_{\delta}^{1/2} \frac{G(t)}{t^3}\, dt.$$

The first and third terms tend to 0 as $n \longrightarrow \infty$. Since $(1/t)G(t) < \epsilon$ for $|t| \leq \delta$ the second term is dominated by

$$\frac{2A\epsilon}{n+1} \int_{1/(n+1)}^{\delta} \frac{dt}{t^2} < 2A\epsilon.$$

Thus, $|\sigma_n(x) - f(x)|$ can be made as small as we wish by choosing n large enough. This proves the theorem of Lebesgue.

An application of this theorem is that we can reverse the inequality in Bessel's inequality (see (1.5)). For, if $f \in L^2(0,1) \subset L^1(0,1)$ and $c_k = \hat{f}(k)$, $k = 0, \pm 1, \pm 2, \cdots$, then the $(C, 1)$ means of f have the form

$$\sigma_n(x) = \sum_{k=-n}^{n} \left(1 - \frac{|k|}{n+1}\right) c_k e^{2\pi i k x}.$$

Using the orthogonality relations (1.3) we have

$$\int_0^1 |\sigma_n(x)|^2\, dx = \sum_{k=-n}^{n} \left(1 - \frac{|k|}{n+1}\right)^2 |c_k|^2 \leq \sum_{k=-\infty}^{\infty} |c_k|^2.$$

Since $\sigma_n(x) \longrightarrow f(x)$ almost everywhere, Fatou's lemma implies that

$$\int_0^1 |f(x)|^2\, dx \leq \varliminf_{n \to \infty} \int_0^1 |\sigma_n(x)|^2\, dx.$$

Consequently,

$$\int_0^1 |f(x)|^2\, dx \leq \sum_{k=-\infty}^{\infty} |c_k|^2.$$

Together with Bessel's inequality this gives us the following relation, known as *Parseval's formula:*

$$\int_0^1 |f(x)|^2\, dx = \sum_{k=-\infty}^{\infty} |\hat{f}(k)|^2. \tag{3.7}$$

We can also show easily that the partial sums of the Fourier series of a function f in $L^2(0, 1)$ converge to f in the L^2-norm. We have already seen that they do converge to a function g in $L^2(0, 1)$ (see the argument preceding (1.5)). This implies that g and f have the same Fourier coefficients $\{c_k\} = \{\hat{f}(k)\}$; for

$$\int_0^1 g(t)e^{-2\pi ikt}\,dt = \int_0^1 [g(t) - s_n(t)]e^{-2\pi ikt}\,dt + \int_0^1 s_n(t)e^{-2\pi ikt}\,dt.$$

The first term of this sum is dominated, in absolute value, by $||g - s_n||_2$ (use Schwarz's inequality) and, thus, tends to 0 as $n \longrightarrow \infty$. The second term equals c_k as long as $n \geq k$. From this we conclude that the Fourier coefficients of the function $f - g$ are all 0. But, since the system $\{e^{2\pi inx}\}$ is complete, this implies $f(x) - g(x) = 0$ almost everywhere.

Let us observe that if we had started with a square summable sequence $\{c_k\}$ (that is, $\sum_{k=-\infty}^{\infty} |c_k|^2 < \infty$), then, by the orthogonality relations (1.3), the partial sums $s_n(x)$ of $\sum_{k=-\infty}^{\infty} c_k e^{2\pi ikx}$ converge in the L^2-norm to a function g. The argument just used shows that $\hat{g}(k) = c_k$ for $k = 0, \pm 1, \pm 2, \cdots$.

We collect these facts together in the following statement:

(3.8) *Suppose f belongs to* $L^2(0, 1)$ *then its Fourier series converges to f in the L^2-norm; that is,*

$$||f - s_n||_2 = \left(\int_0^1 |f(x) - s_n(x)|^2\,dx\right)^{1/2} = \left(\int_0^1 \left| f(x) - \sum_{k=-n}^{n} \hat{f}(k)e^{2\pi ikx}\right|^2 dx\right)^{1/2}$$

tends to 0 as n tends to ∞. Furthermore,

$$||f||_2 = \left(\int_0^1 |f(x)|^2\,dx\right)^{1/2} = \left(\sum_{k=-\infty}^{\infty} |\hat{f}(k)|^2\right)^{1/2} = ||\hat{f}||_2.$$

If a sequence $\{c_k\}$ satisfies $\sum_{k=-\infty}^{\infty} |c_k|^2 < \infty$ then there exists a function f in $L^2(0, 1)$ such that $c_k = \hat{f}(k)$ for all integers k.

Except for not having proved (1.7) in the case of Abel summability we have now established all the results announced in the first section in connection with the Fourier inversion problem. We leave it to the reader to show that essentially the argument used above for Cesàro summability, using the estimates

(a′) $P(r, t) \leq \frac{1}{1-r}$ and

(b′) $P(r, t) \leq \frac{A(1-r)}{t^2}$, $|t| \leq \frac{1}{2}$, $0 \leq r < 1$, where A is an absolute constant,

gives us (1.7) in full.

As we stated toward the end of the last section, we shall not essentially repeat all this material by giving the corresponding results in the theory of Fourier integrals. The reader should have no trouble, for example, in stating and proving the analogs of results (1.6) and (1.7). Nevertheless, some discussion of what happens when we carry over the above material to the case of the real line is in order.

First of all, we cannot adapt the argument we gave to establish the Riemann-Lebesgue theorem to the case of functions in $L^1(-\infty, \infty)$. For one thing, we have not even defined the Fourier transform for functions in $L^2(-\infty, \infty)$; moreover, as we shall see shortly after we define it, it is *not* true in general that $\hat{f}(x) \longrightarrow 0$ as $|x| \longrightarrow \infty$ when $f \in L^2(-\infty, \infty)$. We shall show, however, that the Riemann-Lebesgue theorem does extend to the case of the real line, and the simple argument we shall give can be adapted to prove (3.1) as well. We shall prove

(3.9) *If* $f \in L^1(-\infty, \infty)$ *then* $\hat{f}(x) \longrightarrow 0$ *as* $|x| \longrightarrow \infty$.

Since

$$
\begin{aligned}
-\hat{f}(x) &= \int_{-\infty}^{\infty} (-1)e^{-2\pi ixt} f(t)\, dt \\
&= \int_{-\infty}^{\infty} e^{-2\pi ix[t-(1/2x)]} f(t)\, dt \\
&= \int_{-\infty}^{\infty} e^{-2\pi ixt} f\left(t + \frac{1}{2x}\right) dt,
\end{aligned}
$$

we have

$$|\hat{f}(x)| = \left|\frac{1}{2}\int_{-\infty}^{\infty}\left\{f(t) - f\left(t + \frac{1}{2x}\right)\right\} e^{-2\pi ixt}\, dt\right|$$

$$\leq \frac{1}{2}\,\omega_1\left(\frac{1}{2x}\right) \longrightarrow 0 \qquad \text{as } x \longrightarrow \infty.$$

Let us now state the result that corresponds to (3.8):

(3.10) THE PLANCHEREL THEOREM: *If f belongs to $L^2(-\infty, \infty)$ then there exists a function $\hat{f}$, also in $L^2(-\infty, \infty)$, such that*

$$\int_{-\infty}^{\infty}\left|\hat{f}(x) - \int_{-N}^{N} e^{-2\pi ixt} f(t)\, dt\right|^2 dx \longrightarrow 0$$

as $N \longrightarrow \infty$. The function $\hat{f}$ is called the Fourier transform of f and it agrees a.e. with the previously defined Fourier transform whenever $f \in L^1(-\infty, \infty) \cap L^2(-\infty, \infty)$. Furthermore, Parseval's formula holds

$$\|\hat{f}\|_2 = \|f\|_2.$$

Fourier inversion is possible in the L^2-norm:

$$\int_{-\infty}^{\infty}\left|f(t) - \int_{-N}^{N} e^{2\pi ixt}\hat{f}(x)\, dx\right|^2 dt \longrightarrow 0$$

as $N \longrightarrow \infty$. Finally, each f in $L^2(-\infty, \infty)$ has the form $f = \hat{g}$ for an (almost everywhere) unique g in $L^2(-\infty, \infty)$.

To prove (3.10) let us choose an f in $L^1(-\infty, \infty) \cap L^2(-\infty, \infty)$ and form the convolution $h = f * g$ where $g(t) = \overline{f(-t)}$. It follows from the remarks made at the very end of Sec. 2 that h, being the convolution of two functions in $L^1(-\infty, \infty)$, is integrable and, also, $\hat{h}(x) = \hat{f}(x) \cdot \overline{\hat{f}}(x) = |\hat{f}(x)|^2 \geq 0$ (by the real-line analog of (1.18)). Moreover, we claim that 0 is a point of the Lebesgue set of h; in fact, it is a point of continuity of h. For

$$|h(\delta) - h(0)| = \left|\int_{-\infty}^{\infty}\{g(\delta - t) - g(-t)\} f(t)\, dt\right|$$

$$\leq \left[\int_{-\infty}^{\infty}|g(\delta - t) - g(-t)|^2\, dt\right]^{1/2}\|f\|_2.$$

But $\left[\int_{-\infty}^{\infty}|g(\delta - t) - g(-t)|^2\, dt\right]^{1/2}$ is dominated by the L^2 modulus of continuity evaluated at δ, $\omega_2(\delta)$, of the function whose value at t is $g(-t)$. Since the latter belongs to $L^2(-\infty, \infty)$, we can conclude that $\lim_{\delta \to 0} |h(\delta) - h(0)| = 0$. Consequently, the $(C, 1)$ means of the integral $\int_{-\infty}^{\infty} \hat{h}(x) e^{2\pi i x t}\, dx$, defining $(\hat{h})^{\vee}(x)$, converge to $h(t)$ when $t = 0$. That is,

$$\int_{-R}^{R} \left(1 - \frac{|x|}{R}\right) \hat{h}(x)\, dx \longrightarrow h(0).$$

But $\hat{h}(x) \geq 0$ and the integrand in this integral increases monotonically to $\hat{h}(x)$. Thus, by the Lebesgue monotone convergence theorem $\hat{h}$ is integrable and

$$\int_{-\infty}^{\infty} \hat{h}(x)\, dx = h(0).$$

Since $h(0) = \int_{-\infty}^{\infty} f(t)\overline{f(t)}\, dt = ||f||_2^2$ and $\hat{h} = |\hat{f}|^2$ this shows

$$\int_{-\infty}^{\infty} |\hat{f}(x)|^2\, dx = \int_{-\infty}^{\infty} |f(x)|^2\, dx. \tag{3.11}$$

Thus, Parseval's formula holds when $f \in L^1(-\infty, \infty) \cap L^2(-\infty, \infty)$.

In particular, (3.11) tells us that the mapping $f \longrightarrow \hat{f}$ is bounded, in the L^2-norm, as a linear operator on the dense subset $L^1 \cap L^2$ of the Hilbert space L^2 into L^2. It is well known that in such a case there exists a unique, bounded extension of the operator on all of the Hilbert space. Using the same notation for this extension we then can conclude that (3.11) holds for all $f \in L^2(-\infty, \infty)$.

If we let χ_N be the characteristic function of the interval $[-N, N]$ we set $f_N = \chi_N f$, for $f \in L^2(-\infty, \infty)$. Then $f_N \in L^1(-\infty, \infty) \cap L^2(-\infty, \infty)$ and $||f - f_N||_2 \longrightarrow 0$ as $N \longrightarrow \infty$. Because of the boundedness of the operator we have just defined we then must have $||\hat{f} - \hat{f}_N||_2 \longrightarrow 0$. This proves the first part of (3.10).

The Fourier inversion part of Plancherel's theorem follows easily from the relation

$$\int_{-\infty}^{\infty} f(x)\hat{g}(x)\, dx = \int_{-\infty}^{\infty} \hat{f}(x) g(x)\, dx \tag{3.12}$$

whenever $f, g \in L^2(-\infty, \infty)$. The proof of (3.12) for functions in $L^1 \cap L^2$ is straightforward; thus, we first establish (3.12) for f_N and g_N and obtain the general result by letting $N \longrightarrow \infty$. It is an immediate consequence of (3.12) that

$$\bar{f} = (\hat{\bar{f}})^\wedge \tag{3.13}$$

for all $f \in L^2(-\infty, \infty)$. For

$$\|\bar{f} - (\hat{\bar{f}})^\wedge\|_2^2 = \left\{\int f\bar{f} - \int f(\hat{\bar{f}})^\wedge\right\} - \overline{\left\{\int f(\hat{\bar{f}})^\wedge - \int (\hat{\bar{f}})^\wedge\overline{(\hat{\bar{f}})^\wedge}\right\}}.$$

But both expressions in the brackets are 0; applying (3.12) to the first one we obtain $\int f\bar{f} - \int \hat{f}\bar{\hat{f}}$, which is 0 by Parseval's formula. A similar argument shows the second expression is 0 also. Now, because of (3.13) we have that $\bar{f}$ is the limit in the L^2-norm, as $N \longrightarrow \infty$, of the functions given by the integrals

$$\int_{-N}^{N} e^{-2\pi ixt}\,\bar{\hat{f}}(t)\,dt = \overline{\int_{-N}^{N} e^{2\pi ixt}\hat{f}(t)\,dt}.$$

By taking complex conjugates we have the Fourier inversion result announced in (3.10).

The last statement follows from this inversion applied to $g = f$, where $\check{f}$ is the limit in L^2, as $N \longrightarrow \infty$, of

$$\int_{-N}^{N} e^{2\pi ixt}f(t)\,dt = \overline{(\bar{f}_N)^\wedge(x)}.$$

The material presented up to this point belongs to the foundation of the theories of Fourier series and integrals. It is desirable also to describe some of the directions in which harmonic analysis has been developed. This subject, however, is so rich with results, covering such a wide field of mathematics, that it is impossible to present something that approximates a survey of the highlights in the space that we have available. For this reason we shall, from time to time, select certain topics and present them, mostly without proofs. The bases for our selections are the possibility of extending these topics to other parts of harmonic analysis, the simplicity of the concepts involved, and their applicability to other branches of mathematics. We conclude this section with what is perhaps the one result that best fits this description, a celebrated theorem of Wiener. As we shall see in Sec. 5, this theorem extends to the harmonic analysis associated with any locally compact abelian group, no new concepts are needed to understand it, and from it one can prove rather easily the prime number theorem [4, p. 303]!

Suppose f belongs to $L^1(-\infty, \infty)$ then the collection of all finite linear combinations of translates of f will be denoted by T_f. That is, g belongs to T_f if and only if it has the form

$$g(x) = \Sigma\, a_k f(x + t_k) \tag{3.14}$$

for some finite set of real numbers t_k and complex numbers a_k. The theorem of Wiener asserts the following:

(3.15) *Suppose f belongs to $L^1(-\infty, \infty)$ and that $\hat{f}(x)$ is never* 0, *then the closure, in the L^1 topology, of T_f is all of $L^1(-\infty, \infty)$. In other words, any function in $L^1(-\infty, \infty)$ can be approximated arbitrarily closely in the L^1-norm by functions of the form* (3.14).

It is easy to find functions f whose Fourier transforms never vanish. For example, the proof of (2.16), with $y = 1$, consisted, simply, of showing that when $f(t) = e^{-2\pi|t|}$ then $\hat{f}(x) = \dfrac{1}{\pi}\dfrac{1}{1 + x^2}$.

That the condition $\hat{f}(x) \neq 0$ for all real x is necessary is clear. For, if $\hat{f}(x_0) = 0$ for some x_0 and $g \in T_f$ then g has the form (3.14) and, thus,

$$\hat{g}(x) = \Sigma\, a_k e^{2\pi i t_k} \hat{f}(x).$$

Therefore $\hat{g}(x_0) = 0$. Now suppose h is in the L^1 closure of T_f; then there exists a sequence $\{g_n\}$ in T_f such that $\|g_n - h\|_1 \longrightarrow 0$ as $n \longrightarrow \infty$. Thus, by (2.9),

$$|\hat{g}_n(x_0) - \hat{h}(x_0)| \leq \|\hat{g}_n - \hat{h}\|_\infty \leq \|g_n - h\|_1 \longrightarrow 0 \quad \text{as } n \longrightarrow \infty.$$

Since $g_n(x_0) = 0$ for all n we must have $\hat{h}(x_0) = 0$. We have shown that the Fourier transforms of all h in the closure of T_f vanish at x_0. Since there are integrable functions whose Fourier transforms never vanish, this closure cannot comprise all of $L^1(-\infty, \infty)$.

We shall not give a proof of (3.15). We would like to point out, however, that this proof uses strongly the fact that whenever $f \in L^1(-\infty, \infty)$ then the closure of T_f is a (closed) *ideal* in

$L^1(-\infty, \infty)$. By the term "ideal" we mean a linear subspace, I, of $L^1(-\infty, \infty)$ such that $g * h \in I$ whenever $g \in I$ and $h \in L^1(-\infty, \infty)$.

An important class of ideals in $L^1(-\infty, \infty)$ is the collection of closed *maximal ideals* (an ideal M is said to be maximal if it is not contained in any proper ideal in $L^1(-\infty, \infty)$ other than M itself). This class has a very elegant characterization:

(3.16) *M is a closed maximal ideal if and only if there exists a real number x such that M consists of all $f \in L^1(-\infty, \infty)$ such that $\hat{f}(x) = 0$.*

Thus, we have a one-to-one correspondence between the real numbers and the closed maximal ideals in $L^1(-\infty, \infty)$. We shall denote the closed maximal ideal corresponding to x by $M(x)$. It is not hard to see that the following is a generalization of Wiener's theorem:

(3.15′) *Every proper closed ideal in $L^1(-\infty, \infty)$ is contained in a closed maximal ideal.*

For if $\hat{f}(x)$ is never 0 then f cannot belong to $M(x)$ for any x. Thus, the closed ideal obtained by taking the closure of T_f cannot be included in any closed maximal ideal. Consequently, this ideal is not proper; that is, it must coincide with $L^1(-\infty, \infty)$.

These considerations lead us to the formulation of a well-known problem in harmonic analysis, the problem of *spectral synthesis.* If the closure of T_f is a proper ideal, I_f, then, by (3.15′), it is contained in a certain class of ideals $M(x)$. It is easy to check that the intersection of all closed maximal ideals containing I_f is a closed ideal. The problem of spectral synthesis is to determine for which $f \in L^1(-\infty, \infty)$ it is true that I_f equals this intersection. It has been discovered only recently (in 1959) that there are $f \in L^1(-\infty, \infty)$ for which I_f is not equal to the intersection of all maximal ideals containing it.

It is often useful to rephrase this problem in the following way: *For which f in $L^1(-\infty, \infty)$ is it true that if $\hat{g}(x) = 0$ whenever $\hat{f}(x) = 0$ then g is in the closure of T_f?*

4. SOME OPERATORS THAT ARISE IN HARMONIC ANALYSIS

Suppose $F(z) = a_0 + a_1 z + a_2 z^2 + \cdots + a_n z^n + \cdots$ is an analytic function in the interior of the unit circle. Suppose, further, that F is bounded in this domain; say, $|F(z)| \leq B < \infty$ for $|z| < 1$. Let us write $z = re^{2\pi i\theta}$, $0 \leq r < 1$, $0 \leq \theta < 1$. Then, using the orthogonality relations (1.3),

$$\sum_{k=0}^{\infty} |a_k|^2 r^{2k} = \int_0^1 \left(\sum_{k=0}^{\infty} a_k r^k e^{2\pi ik\theta}\right)\left(\sum_{k=0}^{\infty} \bar{a}_k r^k e^{-2\pi ik\theta}\right) d\theta$$

$$= \int_0^1 |F(re^{2\pi i\theta})|^2 \, d\theta \leq B^2 \qquad \text{for } 0 \leq r < 1.$$

Letting $r \longrightarrow 1$ we therefore obtain $\sum_{k=0}^{\infty} |a_k|^2 < \infty$. By (3.8) we thus can conclude that there exists an f belonging to $L^2(0, 1)$ such that $\hat{f}(k) = a_k$, $k = 0, 1, 2, \cdots$ and $\hat{f}(k) = 0$ for all negative integers k. This shows that

$$F(re^{2\pi i\theta}) = \sum_{k=0}^{\infty} \hat{f}(k) r^k e^{2\pi ik\theta} = \sum_{k=-\infty}^{\infty} \hat{f}(k) r^k e^{2\pi ik\theta}, \qquad 0 \leq r < 1,$$

are the Abel means of the Fourier series of f. By (1.7), therefore, $\lim_{r\to 1} F(re^{2\pi i\theta}) = f(\theta)$ for almost every θ. In particular, we have proved

(4.1) FATOU'S THEOREM: *If F is a bounded analytic function in the interior of the unit circle then the radial limits $\lim_{r\to 1} F(re^{2\pi i\theta})$ exist for almost every θ in $[0, 1)$.*

We shall use this theorem to define an important operator, the *conjugate function mapping*, acting on integrable and periodic functions. Suppose f is such a function. It follows from our discussion concerning the Poisson kernel and the conjugate Poisson kernel that the function G defined by

$$G(z) = \int_0^1 \frac{1 + re^{2\pi i(\theta - t)}}{1 - re^{2\pi i(\theta - t)}} f(t) \, dt \tag{4.2}$$

$$= \int_0^1 P(r, \theta - t) f(t) \, dt + i \int_0^1 Q(r, \theta - t) f(t) \, dt,$$

$z = re^{2\pi i\theta}$, is analytic in the interior of the unit circle. We already know that the first expression in the last sum has radial limits, as $r \longrightarrow 1$, for almost all θ. The following theorem asserts that this is also true for the second term.

(4.3) *Suppose* $f \in L^1(0, 1)$; *then the limits*, $\tilde{f}(\theta)$, *as* $r \longrightarrow 1$, *of*

$$\tilde{A}(r, \theta) = \int_0^1 Q(r, \theta - t) f(t)\, dt$$

$$= \int_0^1 \frac{2r \sin 2\pi(\theta - t)}{1 - 2r \cos 2\pi(\theta - t) + r^2} f(t)\, dt$$

exist for almost all θ. *The function* $\tilde{f}$ *is called the conjugate function of* f.†

By decomposing f into its real and imaginary parts and considering separately the positive and negative parts of each of these, we see that it suffices to prove (4.3) for $f \geq 0$. Thus, letting $A(r, \theta)$ be the Poisson integral and $\tilde{A}(r, \theta)$ the conjugate Poisson integral of f, we obtain an analytic function for $|z| < 1$, $z = re^{2\pi i\theta}$, and its values lie in the right half-plane (by property (B′) of the Poisson kernel). Thus,

$$F(z) = e^{-A(r,\theta) - i\tilde{A}(r,\theta)}$$

† If we let $r \longrightarrow 1$ we obtain, formally,

$$\tilde{f}(\theta) = \int_0^1 \frac{2 \sin 2\pi(\theta - t)}{2(1 - \cos 2\pi(\theta - t))} f(t)\, dt = \int_0^1 \frac{f(t)}{4 \tan \pi(\theta - t)}\, dt.$$

This last integral, however, is not defined even when f is an extremely well-behaved function (for example, if f is constant and nonzero in a neighborhood of θ the integral fails to exist). One can show, however, that if we take the principal value integral

$$(*) \qquad \lim_{\epsilon \to 0+} \int_{\substack{\epsilon \leq |\theta - t| \\ 0 \leq t \leq 1}} \frac{f(t)}{4 \tan \pi(\theta - t)}\, dt$$

we do obtain a value for almost all θ. In fact, an argument not unlike that used to prove (1.7) shows that the existence of these limits, as $\epsilon = 1 - r \longrightarrow 0$, is equivalent almost everywhere to the existence of the limits of (4.3). One may take (*) as the definition of the conjugate function, therefore, and avoid the use of analytic function theory. However, the real-variable proof of the existence of $\tilde{f}$ is by no means easy.

is a bounded ($|F(z)| \leq 1$) analytic function in the interior of the unit circle. By Fatou's theorem (4.1) the radial limits of F exist almost everywhere. Since the radial limits of $A(r, \theta)$ also exist almost everywhere and are finite (they equal $f(\theta)$), the limits of F must be nonzero almost everywhere. But this implies the existence of $\lim_{r \to 1} \tilde{A}(r, \theta)$ for almost all θ, and (4.3) is proved.

The conjugate function mapping is obviously linear. If $f \in L^2(0, 1)$, then, using the fact that $\tilde{A}(r, \theta)$, $0 \leq r < 1$, $0 \leq \theta < 1$, are the Abel means of the conjugate Fourier series of f and, also, the result (3.8), we can show very easily that

$$\text{(4.4)} \qquad \|\tilde{f}\|_2^2 = \sum_{|k| \geq 1} |\hat{f}(k)|^2 \leq \sum_{k=-\infty}^{\infty} |\hat{f}(k)|^2 = \|f\|_2^2.$$

Thus the mapping $f \longrightarrow \tilde{f}$ is a bounded linear transformation when restricted to the space $L^2(0, 1)$. One can show, however, that there are functions in $L^1(0, 1)$ for which the conjugate function is not integrable. In particular, it follows that this mapping is not bounded as an operator from $L^1(0, 1)$ into $L^1(0, 1)$. We do have the following theorem, however.

(4.5) THEOREM OF M. RIESZ: *If* $f \in L^p(0, 1)$, $1 < p < \infty$, *then* $\tilde{f} \in L^p(0, 1)$ *and*

$$\|\tilde{f}\|_p \leq A_p \|f\|_p,$$

where A_p *depends only on* p.

For exactly the same reasons that we gave in the proof of (4.3) it suffices to consider the case $f \geq 0$. Furthermore, we claim that it is sufficient to show that

$$\text{(4.6)} \qquad \int_0^1 |\tilde{A}(r, \theta)|^p \, d\theta \leq c_p \int_0^1 |A(r, \theta)|^p \, d\theta = c_p \int_0^1 [A(r, \theta)]^p \, d\theta$$

for $0 \leq r < 1$, where c_p depends only on p. For, an argument very similar to that used to prove (2.14), shows that the Poisson integrals $A(r, \theta)$ converge to $f(\theta)$ in the $L^p(0, 1)$ norms. In particular,

$$\lim_{r \to 1} \int_0^1 [A(r, \theta)]^p \, d\theta = \|f\|_p^p.$$

Since $\tilde{A}(r, \theta) \longrightarrow \tilde{f}(\theta)$ almost everywhere as $r \longrightarrow 1$, an application of Fatou's lemma then gives us the inequality $||\tilde{f}||_p^p \leq c_p||f||_p^p$, from which the theorem follows.

To show (4.6) we argue in the following manner. Let

$$F(z) = A(r, \theta) + i\tilde{A}(r, \theta) \qquad \text{for } x + iy = z = re^{2\pi i\theta}, 0 \leq r < 1,$$

and let $\Delta = \partial^2/\partial x^2 + \partial^2/\partial y^2$ denote the Laplacian operator. Treating A, $\tilde{A}$, and F as functions of x and y we have, by a simple calculation which uses the Cauchy-Riemann equations (recall that F is analytic),

$$\Delta A^p = p(p-1)A^{p-2}|F'|^2 \qquad \text{and} \qquad \Delta|F|^p = p^2|F|^{p-2}|F'|^2.$$

Let us assume, first, that $1 < p \leq 2$. Then, since $|F| \geq A$,

$$\Delta|F|^p \leq q\,\Delta A^p,$$

where $(1/q) = 1 - (1/p)$. We claim that this inequality and Green's formula imply

$$\int_0^1 |F(re^{2\pi i\theta})|^p\, d\theta \leq q \int_0^1 [A(r, \theta)]^p\, d\theta$$

for $0 \leq r < 1$, which certainly implies (4.6). The form of Green's formula we need is the following. Suppose u is a continuous function defined in the unit circle which has continuous first and second derivatives, S is the circle $\{(x, y);\ x^2 + y^2 \leq r^2 < 1\}$ and C its circumference. Then

$$\int_C \frac{\partial u}{\partial r}\, ds = \iint_S \Delta u\, dx\, dy,$$

where $\partial/\partial r$ denotes differentiation in the direction of the radius vector and $ds = r\, d\theta$. Applying this formula to $u = A^p$ and $u = |F|^p$ we obtain, because of the inequality $\Delta|F|^p \leq q\,\Delta A^p$,

$$\int_0^1 \left(\frac{\partial}{\partial r} |F(re^{2\pi i\theta})|^p\right) r\, d\theta \leq q \int_0^1 \left(\frac{\partial}{\partial r} [A(r, \theta)]^p\right) r\, d\theta.$$

Thus, because of the smoothness of the functions involved,

$$\frac{d}{dr} \int_0^1 |F(re^{2\pi i\theta})|^p\, d\theta \leq \frac{d}{dr}\, q \int_0^1 A[(r, \theta)]^p\, d\theta.$$

Since $F(0) = A(0, \theta)$ we obtain the desired inequality by integrating with respect to r.

It remains for us to show that the theorem holds for $p \geq 2$. But it is an easy exercise, using the fact that L^p and L^q are dual when $1/p + 1/q = 1$, to show that whenever a bounded operator acting on L^p is given by a convolution, then it is defined on L^q and is a bounded operator on this space as well. The mapping $f \longrightarrow \tilde{A}(r, \theta)$ is such an operator and it satisfies

$$\int_0^1 |\tilde{A}(r, \theta)|^p \, d\theta \leq B_p \int_0^1 |A(r, \theta)|^p \, d\theta \leq B_p \|f\|_p^p \dagger$$

for $1 < p \leq 2$. Thus, it satisfies this inequality for the indices conjugate to p; that is, for p replaced by $q = p/(p-1)$:

$$\int_0^1 |\tilde{A}(r, \theta)|^q \, d\theta \leq C_q \|f\|_q^q$$

whenever $f \in L^q(0, 1)$, $q \geq 2$. But, by Fatou's lemma, this implies

$$\int_0^1 |\tilde{f}(\theta)|^q \, d\theta = \int_0^1 \lim_{r \to 1} |\tilde{A}(r, \theta)|^q \, d\theta \leq C_q \|f\|_q^q$$

and (4.5) is proved.

This development gives us a glimpse of the role that "complex methods" (that is, the use of the theory of analytic functions of a complex variable) play in the theory of Fourier series.

Let us examine some more operators that arise naturally in harmonic analysis. For example, let us study the Fourier transform mapping acting on functions defined on the entire real line. Inequality (2.9) tells us that it is a bounded transformation defined on $L^1(-\infty, \infty)$ with values in $L^\infty(-\infty, \infty)$. The Plancherel theorem (3.10) tells us that it is a bounded transformation from $L^2(-\infty, \infty)$ into itself. A natural question, then, is whether it can be defined on other classes L^p and, if so, whether we obtain a

---

† One can give several direct proofs of the inequality

$$\int_0^1 |A(r, \theta)|^p \, d\theta \leq \int_0^1 |f(\theta)|^p \, d\theta, \; p \geq 1.$$

Since it is an immediate consequence of Young's inequality (4.8) (for $g(\theta) = P(r, \theta)$ defines a function in L^1 and $A(r, \theta) = (g * f)(\theta)$) we will not prove it here.

bounded transformation with values in some classes L^q. But any function in L^p, $1 < p < 2$, can be written as a sum of a function in L^1 and one in L^2: put $f = f_1 + f_2$, where $f_2(x) = f(x)$ when $|f(x)| \leq 1$ and $f_2(x) = 0$ otherwise; then $f_1 \in L^1$ and $f_2 \in L^2$. Thus, we can write $\hat{f} = \hat{f}_1 + \hat{f}_2$, where $\hat{f}_1$ is defined as the Fourier transform of a function in L^1 while $\hat{f}_2$ is defined by (3.10). The fact that these two definitions agree when a function belongs to $L^1 \cap L^2$ implies that $\hat{f}$ is well defined. The following theorem tells us that Fourier transformation defined on $L^p(-\infty, \infty)$, $1 < p < 2$, is bounded as a mapping into $L^q(-\infty, \infty)$, where q is the conjugate index to p.

(4.7) THE HAUSDORFF-YOUNG THEOREM: *If* $f \in L^p(-\infty, \infty)$, $1 \leq p \leq 2$, *then* $\hat{f} \in L^q(-\infty, \infty)$, *where* $1/p + 1/q = 1$, *and*

$$||\hat{f}||_q \leq ||f||_p.$$

We shall not prove (4.7) immediately. Instead, we shall give examples of some other inequalities that occur in harmonic analysis and then state some general results from which all these inequalities, including (4.5) and (4.7), follow as relatively easy consequences.

(4.8) YOUNG'S THEOREM: *Suppose* $\frac{1}{r} = \frac{1}{p} + \frac{1}{q} - 1$, *where* $\frac{1}{p} + \frac{1}{q} \geq 1$. *If* $f \in L^p(-\infty, \infty)$ *and* $g \in L^q(-\infty, \infty)$ *then* $f * g$ *belongs to* $L^r(-\infty, \infty)$ *and*

$$||f * g||_r \leq ||f||_p ||g||_q.$$

The same result holds for periodic functions if we replace the interval $(-\infty, \infty)$ *by the interval* $(0, 1)$.

The operator on functions defined on $(-\infty, \infty)$ that corresponds to the conjugate function operator satisfies the same inequality (4.5). Using (2.18) and arguments that are completely analogous to those we gave at the beginning of this section, we see that this operator, called the *Hilbert transform*, can be defined by letting

$$\tilde{f}(x) = \lim_{y \to 0+} \tilde{f}(x, y) = \lim_{y \to 0+} \frac{1}{\pi} \int_{-\infty}^{\infty} f(t) \frac{x - t}{(x - t)^2 + y^2} \, dt$$

correspond to $f \in L^p(-\infty, \infty)$, $1 \leq p$, and that the following result holds:

(4.9) *If $f \in L^p(-\infty, \infty)$, $1 < p < \infty$, then its Hilbert transform $\tilde{f}$ also belongs to $L^p(-\infty, \infty)$ and*

$$\|\tilde{f}\|_p \leq A_p \|f\|_p,$$

where A_p depends only on p.

All the operators we have encountered up to this point are linear. There are several important transformations in harmonic analysis, however, that are not linear. Perhaps the best-known example of such a transformation is the *Hardy-Littlewood maximal* function. This operator is defined in the following way: if $f \in L^p(-\infty, \infty)$, $1 \leq p \leq \infty$, then its maximal function is the function whose value at $x \in (-\infty, \infty)$ is

$$f^*(x) = \sup_{h \neq 0} \frac{1}{h} \int_x^{x+h} |f(t)|\, dt.$$

Lebesgue's theorem on the differentiation of the integral guarantees that $f^*(x) < \infty$ for almost every x. It can be shown that

(4.10) *If $f \in L^p(-\infty, \infty)$, $1 < p \leq \infty$, then $f^* \in L^p(-\infty, \infty)$ and*

$$\|f^*\|_p \leq A_p \|f\|_p,$$

where A_p depends only on p.

The usefulness of the maximal function lies in the fact that it majorizes several important operators. Thus, it is clear why a theorem like (4.10) is desirable, as it immediately implies the boundedness of these operators.

Although the mapping $f \longrightarrow f^*$ is not linear, it does satisfy the inequality $(f + g)^* \leq f^* + g^*$. This property is generally referred to as *sublinearity*. More generally, we say that an operator T mapping functions into functions is *sublinear* if, whenever Tf and Tg are defined, so is $T(f + g)$ and

$$|T(f + g)| \leq |Tf| + |Tg|.$$

In all these instances special cases of the inequalities involved are fairly easy to establish. For the conjugate function mapping the case $p = 2$ was seen to be an easy consequence of Theorem

(3.8) (see (4.4)). A similar argument, using the Plancherel theorem, shows that the same is true for the Hilbert transform. We have pointed out that the cases $p = 1$, $q = \infty$ and $p = 2 = q$ of the Hausdorff-Young theorem had already been obtained by us in the previous sections. The inequality $||f * g||_1 \leq ||f||_1||g||_1$, which was the first result (property (i)) we established after introducing the operation of convolution, is the special case $r = p = q = 1$ of Young's theorem. Another special case of this theorem that is immediate is obtained when p and q are conjugate indices, $1/p + 1/q = 1$, and, thus, $r = \infty$; for this is simply a consequence of Hölder's inequality. Finally, it is clear that (4.10) holds when $p = \infty$.

It was M. Riesz who first discovered (in 1927) a general principle that asserted, in part, that in a wide variety of inequalities of the type we are discussing, special cases, such as those described in the previous paragraph, imply the general case. In order to state his theorem, known as the *M. Riesz convexity theorem*, we need to establish some notation. Suppose (M, μ) and (N, ν) are two measure spaces, where M and N are the point sets and μ and ν the measures. An operator T mapping measurable functions on M into measurable functions on N is said to be of *type* (p, q) if it is defined on $L^p(M)$ and there exists a constant A, independent of $f \in L^p(M)$, such that

$$(4.11) \qquad ||Tf||_q = \left(\int_N |Tf|^q \, d\nu\right)^{1/q} \leq A \left(\int_M |f|^p \, d\mu\right)^{1/p} = A||f||_p.$$

The least A for which (4.11) holds is called the *bound*, or *norm*, of T. The general principle can then be stated in the following way.

(4.12) THE M. RIESZ CONVEXITY THEOREM: *Suppose a linear operator T is of types (p_0, q_0) and (p_1, q_1), with bounds A_0 and A_1, respectively. Then it is of type (p_t, q_t), with bound $A_t \leq A_0^{1-t}A_1^t$, for $0 \leq t \leq 1$, where*

$$\frac{1}{p_t} = \frac{1-t}{p_0} + \frac{t}{p_1} \quad \textit{and} \quad \frac{1}{q_t} = \frac{1-t}{q_0} + \frac{t}{q_1}.$$

The Hausdorff-Young theorem is an immediate consequence of this result. Since $T: f \longrightarrow \hat{f}$ is of types $(1, \infty)$ and $(2, 2)$, it must be of type $\left(\frac{2}{2-t}, \frac{2}{t}\right)$ for $0 \leq t \leq 1$. But if $p = \frac{2}{2-t}$, then the conjugate index is $q = \frac{p}{p-1} = \frac{2}{t}$. Since the "end-point" (i.e. $t = 0$ and $t = 1$) bounds are 1, we have

$$\|\hat{f}\|_q \leq 1^{(1-t)}1^t\|f\|_p = \|f\|_p,$$

which is the inequality in (4.7).

Similarly, Young's theorem follows from (4.12). First, let us fix $g \in L^1$ and define $Tf = f * g$. We have seen that T is of type $(1, 1)$, with bound $\|g\|_1$, and of type (∞, ∞), also with bound $\|g\|_1$. Thus, T is of type $\left(\frac{1}{1-t}, \frac{1}{1-t}\right)$, $0 \leq t \leq 1$, with a bound less than or equal to $\|g\|_1^{1-t}\|g\|_1^t = \|g\|_1$. Putting $p = \frac{1}{1-t}$ this gives us (4.8) with $r = p$ and $q = 1$. To obtain the general case we fix $f \in L^p$ and define $Tg = f * g$. We have just shown that T is of type $(1, p)$ with bound $\|f\|_p$. Letting $g \in L^q$, where q is conjugate to p, we also have T of type (q, ∞) with bound $\|f\|_p$. Thus, T is of type (p_t, q_t), with bound no greater than $\|f\|_p$, where $p_t = \frac{p}{p-t}$ and $q_t = \frac{p}{1-t}$, $0 \leq t \leq 1$. That is,

$$\|f * g\|_{q_t} \leq \|f\|_p\|g\|_{p_t}.$$

Since it follows immediately that $\frac{1}{q_t} = \frac{1}{p} + \frac{1}{p_t} - 1$, and, as t ranges between 0 and 1, $\frac{1}{p} + \frac{1}{p_t}$ ranges from $\frac{1}{p} + 1$ to 1, this is precisely the inequality of (4.8).

Unfortunately none of the other inequalities we stated can be derived from the special cases discussed above and the M. Riesz convexity theorem. For example, the conjugate function mapping, as we have seen, is easily seen to be of type $(2, 2)$. Were we able to show that it is of type $(1, 1)$ it would then follow that it is of type (p, p), $1 < p < 2$, and this, in turn, would imply the result for $p > 2$ (as we saw at the end of the proof of (4.5)). But we have already stated that this operator is not a bounded transformation on $L^1(0, 1)$. Nevertheless, there is a substitute result, due to Kolmogoroff, and an extension of the M. Riesz convexity theorem, due to Marcinkiewicz, that does allow us to obtain Theorem (4.5) much in the same way we obtained (4.7) and (4.8). Furthermore, this method is applicable to Theorems (4.9) and (4.10) as well.

The substitute result of Kolmogoroff is a condition that is weaker than type (1, 1). We shall consider this condition in a more general setting. First, however, we need to introduce the concept of the distribution function of a measurable function. Let g be a measurable function defined on the measure space (N, ν) and, for $y > 0$, $E_y = \{x \in N; |g(x)| > y\}$. Then the *distribution function of* g is the nonincreasing function $\lambda = \lambda_g$ defined for all $y > 0$ by

$$\lambda(y) = \nu(E_y).$$

It is an easy exercise in measure theory to show that if $g \in L^q(N)$ then

$$(4.13) \qquad ||g||_q = \left(\int_N |g(x)|^q \, d\nu\right)^{1/q} = \left(q \int_0^\infty y^{q-1}\lambda(y) \, dy\right)^{1/q}.$$

Suppose, now, that T is an operator of type (p, q), with bound A, $1 \le q < \infty$, mapping functions defined on M into functions defined on N. Let $f \in L^p(M)$, $g = Tf$, and λ the distribution function of g. Then

$$y^q\lambda(y) = \int_{E_y} y^q \, d\nu \le \int_{E_y} |g(x)|^q \, d\nu \le \int_N |g(x)|^q \, d\nu$$
$$\le \left(A\left[\int_M |f(t)|^p \, d\mu\right]^{1/p}\right)^q.$$

That is,

$$(4.14) \qquad \lambda_g(y) = \lambda(y) \le \left(\frac{A}{y}\,||f||_p\right)^q.$$

This condition is easily seen to be weaker than boundedness. An operator that satisfies (4.14) for all $f \in L^p(M)$ is said to be of *weak-type* (p, q). If $q = \infty$ it is convenient to identify weak-type with type.

Kolmogoroff showed that the conjugate function mapping is of weak-type (1, 1). It is then immediate that the following theorem can be used to prove (4.5):

(4.15) THE MARCINKIEWICZ INTERPOLATION THEOREM: *Suppose T is a sublinear operator of weak types (p_0, q_0) and (p_1, q_1), where $1 \leq p_i \leq q_i \leq \infty$ for $i = 0, 1$, and $q_0 \neq q_1$, $p_0 \neq p_1$. Then T is of type (p, q) whenever*

$$\frac{1}{p} = \frac{1-t}{p_0} + \frac{t}{p_1} \quad \text{and} \quad \frac{1}{q} = \frac{1-t}{q_0} + \frac{t}{q_1},$$

$0 < t < 1$.

Similarly, the *Hilbert transform* can be shown to be of weak type (1, 1); thus (4.9) is also a consequence of (4.15). The same is true of (4.10). We shall not prove any of these facts. The reader, however, should have no difficulty in checking that the maximal function mapping cannot be of type (1, 1) (take for f the characteristic function of a finite interval; then f^* is not integrable). The proof that it is of weak type (1, 1) is not hard. The corresponding results for the conjugate function and for the Hilbert transform, however, are somewhat more difficult.

The M. Riesz convexity theorem and the Marcinkiewicz interpolation theorem have many more applications. The examples discussed in this section, however, are sufficient to illustrate the role they play in harmonic analysis.

5. HARMONIC ANALYSIS ON LOCALLY COMPACT ABELIAN GROUPS

We have discussed harmonic analysis associated with three different domains, the circle group (or the group of reals modulo one), the group of integers, and the (additive) group of real numbers. All of these are examples of *locally compact abelian groups*. These are abelian (commutative) groups G, with elements $x, y, z, \cdots$, endowed with a locally compact Hausdorff topology in such a way that the maps $x \longrightarrow x^{-1}$ and $(x, y) \longrightarrow xy$ (defined on G and $G \times G$, respectively) are continuous (we are following the usual custom of writing the operation on G as multiplication and not as addition—which was the case in our three examples;

this should not be a source of confusion to the reader). In this section we shall indicate how harmonic analysis can be extended to functions defined on such groups.

On each such group G there exists a nontrivial regular measure M that, in analogy with Lebesgue measure, has the property that it is invariant with respect to translation. By this we mean that whenever A is a measurable subset of G then $m(A) = m(Ax)$ for all $x \in G$. This is equivalent to the assertion

$$\int_G f(yx)\,dm(y) = \int_G f(y)\,dm(y) \tag{5.1}$$

for all $x \in G$ whenever f is an integrable function. It is obvious that any constant multiple of m also has this property. Conversely, it can be shown that any regular measure satisfying this invariance property must be a constant multiple of m. Such measures are known as *Haar measures*.

The operation of convolution of two functions f and g in $L^1(G)$ is defined, as in the classical case, by the integral

$$(f * g)(x) = \int_G f(xy^{-1})g(y)\,dy.$$

The four properties (i), (ii), (iii), and (iv) (see the end of the first section) hold in this case as well. In particular, $f * g \in L^1(G)$ and $||f * g||_1 \leq ||f||_1||g||_1$.

Moreover, we shall now show that it is possible to give a definition of the Fourier transform so that (1.18) also holds; that is, $(f * g)^\wedge = \hat{f}\hat{g}$ for all f and g in $L^1(G)$. We have seen that the Fourier transform of f is not usually defined on the domain of f. In case of the circle group, for example, Fourier transformation gave us functions defined on the integers. In order to describe the general situation we shall need the concept of a character: By a *character* of a locally compact group G we mean a continuous function, $\hat{x}$, on G such that $|\hat{x}(x)| = 1$ for all x in G and $\hat{x}(xy) = \hat{x}(x)\hat{x}(y)$ for all $x, y \in G$.

The collection of all characters of G is usually denoted by $\hat{G}$. If we define multiplication in $\hat{G}$ by letting $\hat{x}_1\hat{x}_2(x) = \hat{x}_1(x)\hat{x}_2(x)$, for all $x \in G$, whenever $\hat{x}_1, \hat{x}_2 \in \hat{G}$, $\hat{G}$ then becomes an abelian group. We introduce a topology on $\hat{G}$ by letting the sets

$$U(\epsilon, C, x_0) = \{\hat{x} \in \hat{G}; |\hat{x}(x) - \hat{x}_0(x)| < \epsilon, x \in C\},$$

where $\hat{x}_0 \in \hat{G}$, $\epsilon > 0$, and C is a compact subset of G, form a basis. With this topology $\hat{G}$ is then also a locally compact abelian group. $\hat{G}$ is usually called the *character group of* G or the *dual group of* G.

For example, when G is the group of real numbers we easily see that if we let a be a real number, then the mapping $\hat{x}: x \longrightarrow e^{2\pi iax}$, defined for all real x, is a character. One can show that all characters are of this type. Thus, there is a natural one-to-one correspondence between the group of real numbers and $\hat{G}$. Furthermore, this correspondence is a homeomorphism. Hence, we can identify G with $\hat{G}$ in this case.

If G is the group of reals modulo 1 the mappings $\hat{x}: x \longrightarrow e^{2\pi iax}$, $x \in G$, where a is an integer, are characters, and each character has this form. Thus, $\hat{G}$ and the integers are in a one-to-one correspondence that, in this case also, can be shown to be a homeomorphism. Therefore, we can identify $\hat{G}$ with the integers. Similarly the dual group of the integers can be identified with the group of reals modulo 1.

In general, if we fix an x in G and consider the mapping $\hat{x} \longrightarrow \hat{x}(x)$ we obtain a character on $\hat{G}$. It can be shown that every character has this form and that this correspondence between G and $(\hat{G})^\wedge$ is a homeomorphism. This result is known as the *Pontrjagin duality theorem* and it is usually stated, simply, by writing the equality $G = (\hat{G})^\wedge$. Because of this duality the functional notation $\hat{x}(x)$ is discarded and the symbol

$$\langle x, \hat{x} \rangle$$

is used instead. Thus, $\langle x, \hat{x} \rangle$ may be thought of as the value of the function x at $\hat{x}$, $x(\hat{x})$, as well as the value of $\hat{x}$ at x; these, two values are clearly equal.

It is now clear, if we let ourselves be motivated by our three classical examples of locally compact abelian groups, that a natural definition of the Fourier transform for $f \in L^1(G)$, when G is a general locally compact abelian group, is to let it be the function $\hat{f}$ on $\hat{G}$ given by

$$\hat{f}(\hat{x}) = \int_G f(x)\, \overline{\langle x, \hat{x} \rangle}\, dm(x).$$

Many of the results we presented in the previous section hold in this case as well. For example, $\hat{f}$ is a continuous function on $\hat{G}$; when $\hat{G}$ is not compact the Riemann-Lebesgue theorem holds:

(5.2) *If $\hat{G}$ is not compact, $f \in L^1(G)$, and $\epsilon > 0$, then there exists a compact set $C \subset \hat{G}$ such that $|\hat{f}(\hat{x})| < \epsilon$ if $\hat{x}$ is outside of C.*

The basic relation (1.18) between convolution and Fourier transformation is true in general:

(5.3) *If f and g belong to $L^1(G)$ then $(f * g)^\wedge = \hat{f}\hat{g}$.*

Wiener's theorem (3.15) is still valid:

(5.4) *If $f \in L^1(G)$ and $\hat{f}(\hat{x})$ is never 0 then any $g \in L^1(G)$ can be approximated arbitrarily closely in the L^1-norm by functions of the form*

$$\Sigma\, a_k f(xt_k),$$

where the a_k's are a finite collection of complex numbers and the t_k's belong to G.

The Plancherel theorem also has an analog to this general case:

(5.5) *If we restrict the transformation $f \longrightarrow \hat{f}$ to $L^1(G) \cap L^2(G)$ then the L^2 norms are preserved; that is, $\hat{f} \in L^2(\hat{G})$ and Parseval's formula holds.*

$$||f||_2 = ||\hat{f}||_2.$$

Furthermore, this transformation can be extended to a norm preserving transformation of $L^2(G)$ onto $L^2(G)$.

Harmonic analysis can be generalized still further. For example, locally compact groups that are not abelian are associated with important versions of harmonic analysis (the theory of spherical harmonics is associated with the group of rotations in 3-space). We will not, however, pursue this topic further.

6. A SHORT GUIDE TO THE LITERATURE

So many books and papers have been written in harmonic analysis that no attempt will be made here to give anything like a comprehensive bibliography. Rather, our intention is to give some *very* brief suggestions to the reader who would like to pursue the subject further.

All that has been discussed here concerning Fourier series is contained in A. Zygmund's two-volume *Trigonometric Series* [10]. This scholarly book contains essentially all the important work that has been done on the subject. Anyone seriously interested in classical (or, for that matter, modern) harmonic analysis would do well to become acquainted with it. It is often worthwhile, however, to read a short treatment of a subject when learning it. R. R. Goldberg's *Fourier transforms* [3] does an excellent job of presenting that part of Fourier integral theory that generalizes to locally compact abelian groups. In this book the reader will find a proof of Wiener's theorem and a more thorough discussion of the problem of spectral synthesis. For more comprehensive treatments of Fourier integral theory we refer the reader to S. Bochner's *Lectures on Fourier Integrals* [2] and E. C. Titchmarsh's *The Theory of the Fourier Integral* [8].

The literature dealing with the more abstract forms of harmonic analysis is also very large. Pontrjagin's classic *Topological Groups* [6] is still highly recommended reading. The same is true of A. Weil's *L'intégration dans les groupes topologiques et ses applications* [9]. Two very readable modern works that treat the subject of harmonic analysis on groups are Rudin's *Fourier Analysis on Groups* [7] and *Abstract Harmonic Analysis* by Hewitt and Ross [5]. We also recommend an excellent survey on this subject by J. Braconnier [1].

REFERENCES

1. Braconnier, J., *L'analyse harmonique dans les groupes abéliens*, Monographies de l'Enseignement mathématique, No. 5.
2. Bochner, S., *Lectures on Fourier Integrals*. Princeton, N.J.: Princeton University Press, 1959.
3. Goldberg, R. R., *Fourier Transforms*. New York: Cambridge University Press, 1961.
4. Hardy, G. H., *Divergent Series*. Oxford: Clarendon Press, 1949

5. Hewitt, E., and K. A. Ross, *Abstract Harmonic Analysis.* Berlin: Springer, 1963.
6. Pontrjagin, L., *Topological Groups.* Princeton, N.J.: Princeton University Press, 1946.
7. Rudin, W., *Fourier Analysis on Groups.* New York: Interscience Publishers, 1962.
8. Titchmarsh, E. C., *The Theory of the Fourier Integral.* Oxford: Clarendon Press, 1937.
9. Weil, A., *L'intégration dans les groupes topologiques et ses applications.* Paris: Hermann, 1940.
10. Zygmund, A., *Trigonometric Series,* 2nd ed. Cambridge: Cambridge University Press, 1959, 2 vols.

Note to page 127. Since the first edition of this book appeared, this problem has been solved by L. Carleson, who has shown that the Fourier series of a function in L^2 converges to the function almost everywhere. Carleson's result has been extended by R. A. Hunt to the case of functions in L^p, $1 < p < 2$. See L. Carleson, "On convergence and growth of partial sums of Fourier series," *Acta Math.*, vol. 116 (1966), pp. 135–157, and R. A. Hunt, "On the convergence of Fourier series," *Proceedings of the Conference on Orthogonal Expansions and Their Continuous Analogues* (Southern Illinois University), 1968, pp. 235–255.

TOEPLITZ MATRICES

Harold Widom

1. INTRODUCTION

Otto Toeplitz is one of the few mathematicians who has had his name attached to two distinct mathematical objects. What is especially unusual in the case of Toeplitz is that these objects have exactly the same name: *Toeplitz matrix*.

The more famous Toeplitz matrices are associated with procedures for attaching "sums" to divergent series. We shall not mention them again. As far as we are concerned a Toeplitz matrix is an array of complex numbers

$$\begin{bmatrix} \cdot & \cdot & \cdot & \cdot & \cdot & \cdot \\ \cdot & c_0 & c_{-1} & c_{-2} & c_{-3} & \cdot \\ \cdot & c_1 & c_0 & c_{-1} & c_{-2} & \cdot \\ \cdot & c_2 & c_1 & c_0 & c_{-1} & \cdot \\ \cdot & c_3 & c_2 & c_1 & c_0 & \cdot \\ \cdot & \cdot & \cdot & \cdot & \cdot & \cdot \end{bmatrix}.$$

What distinguishes such a matrix is that each diagonal has equal entries. Thus the main diagonal consists entirely of c_0's, the diagonal above the main diagonal consists entirely of c_{-1}'s, etc. To construct such a matrix you start with a sequence

$$(1) \qquad \cdots, c_{-2}, c_{-1}, c_0, c_1, c_2, \cdots$$

and form the associated matrix (c_{m-n}) which has c_{m-n} as its m,nth entry. The reason we put all the dots in the matrix is that we have purposely not indicated its size. In fact we shall be concerned with Toeplitz matrices of three sizes: doubly infinite matrices in which the indices take all integral values, semi-infinite matrices in which the indices take all nonnegative integral values, and finite matrices.

Toeplitz matrices have received a lot of attention in recent years, partly because of their applications to, among other things, probability theory and numerical analysis. However, they are even more interesting for their own sake, and in this article we shall ignore any question of applicability. First we introduce some notation which will be used throughout. Let $\mathbf{c}$ denote, for short, the sequence (1), and $\mathbf{T_c}$ the corresponding Toeplitz matrix, of whatever size. Let $\mathbf{a}$ and $\mathbf{b}$ denote appropriately sized vectors (a_n) and (b_n), respectively; in the doubly infinite case $-\infty < n < \infty$, in the semi-infinite case $n \geq 0$, and in the finite case $1 \leq n \leq N$. The basic problem is to find the condition on $\mathbf{c}$ which insures that the equation

$$(2) \qquad \mathbf{T_c a} = \mathbf{b}$$

can always be solved uniquely for $\mathbf{a}$. (In the infinite cases $\mathbf{a}$ and $\mathbf{b}$ will be required to be square summable.) This is equivalent to the problem of determining when $\mathbf{T_c}$ is invertible, that is, when there exists a matrix $\mathbf{T_c^{-1}}$ so that

$$(3) \qquad \mathbf{T_c T_c^{-1}} = \mathbf{T_c^{-1} T_c} = \mathbf{I},$$

where $\mathbf{I}$ denotes the identity matrix. The connection between (2) and (3) is well known: $\mathbf{a}$ is the result of applying $\mathbf{T_c^{-1}}$ to $\mathbf{b}$.

The approach to the problem, the ideas and methods involved, depend very much on the size of $\mathbf{T_c}$, and we must consider the cases individually.

2. DOUBLY INFINITE TOEPLITZ MATRICES

Let us first write out Eq. (2). It says

$$\sum_{n=-\infty}^{\infty} c_{m-n}a_n = b_m, \qquad -\infty < m < \infty. \tag{4}$$

Before we try to solve this equation, or system of equations, we had better make sure we know what it means. We are going to restrict ourselves to vectors belonging to the class l^2;

$$l^2 = \{\mathbf{a}\colon \sum_{-\infty}^{\infty} |a_n|^2 < \infty\}.$$

The choice of l^2 rather than some other class is due partly to convenience, partly to necessity. We shall not dwell on the matter. Now given that **a** and **b** belong to l^2 there must be some restriction put on the sequence **c** we start with. For each vector $\mathbf{a} \in l^2$ the series on the left of (4) must converge for each m, and the resulting vector **b** must lie in l^2. If this occurs we shall say that $\mathbf{T_c}$ *operates on* l^2. It will turn out that once we see exactly what this means in terms of **c**, the problem of inverting $\mathbf{T_c}$ or solving (4) will be easily solved.

Let **a** be the vector given by

$$a_n = \begin{cases} 1, & n = 0, \\ 0, & \text{otherwise.} \end{cases}$$

There is no problem about convergence of the left side of (4) since the series has but one nonzero term. We shall then have $b_m = c_m$ for all m. But if we insist, as we do, that $\mathbf{b} \in l^2$ then we must have $\mathbf{c} \in l^2$. Thus a *necessary* condition that $\mathbf{T_c}$ operate on l^2 is that $\sum |c_n|^2 < \infty$.

Conversely, suppose $\mathbf{c} \in l^2$. Then Schwarz's inequality tells us that for each m

$$\sum_{n=-\infty}^{\infty} |c_{m-n}a_n| \leq \left\{\sum_{n=-\infty}^{\infty} |c_{m-n}|^2\right\}^{1/2} \left\{\sum_{n=-\infty}^{\infty} |a_n|^2\right\}^{1/2}$$

$$= \left\{\sum_{n=-\infty}^{\infty} |c_n|^2\right\}^{1/2} \left\{\sum_{n=-\infty}^{\infty} |a_n|^2\right\}^{1/2}.$$

Thus the series on the left of (4) does converge for every $\mathbf{a} \in l^2$, and the resulting vector $\mathbf{b}$ satisfies

$$|b_m|^2 \leq \sum_{n=-\infty}^{\infty} |c_n|^2 \sum_{n=-\infty}^{\infty} |a_n|^2.$$

This is certainly not enough to insure $\Sigma\, |b_n|^2 < \infty$. One can wonder whether some more refined estimation would give the stronger result, but unfortunately (or fortunately, if it is considered undesirable that mathematics be too simple) this is not the case, i.e. it is not true that $\mathbf{c} \in l^2$ implies that $\mathbf{T_c}$ operates on l^2.

Here is a counterexample. Let α be any number satisfying $\frac{1}{2} < \alpha < 1$, and define $\mathbf{c}$ by

$$c_n = \begin{cases} n^{-\alpha}, & n > 0, \\ 0, & \text{otherwise.} \end{cases}$$

Since $\alpha > \frac{1}{2}$ we do have $\mathbf{c} \in l^2$. Next let β be any number satisfying

$$\tfrac{1}{2} < \beta < \tfrac{3}{2} - \alpha$$

(such a β exists since $\alpha < 1$) and set

$$a_n = \begin{cases} n^{-\beta}, & n > 0, \\ 0, & \text{otherwise.} \end{cases}$$

Then $\mathbf{a} \in l^2$ but we shall show that the vector $\mathbf{b}$ defined by (4) does not belong to l^2. For $m > 0$ we have

$$b_m = \sum_{n=-\infty}^{\infty} c_{m-n} a_n = \sum_{n=0}^{m-1} (m-n)^{-\alpha} n^{-\beta}.$$

Now let us consider the sum, not all the way to $m - 1$, but only to $[m/2]$, the greatest integer in $m/2$. For $0 \leq n \leq [m/2]$ the term $(m - n)^{-\alpha}$ is at least $m^{-\alpha}$ and the term $n^{-\beta}$ is at least $(m/2)^{-\beta}$. It follows that b_m is at least $m^{1-\alpha-\beta}$ times some positive constant, and since $\alpha + \beta < \frac{3}{2}$ this implies $\mathbf{b} \notin l^2$.

Well, if $\mathbf{c} \in l^2$ is not both necessary and sufficient for $\mathbf{T_c}$ to operate on l^2, what is? It turns out that there is no simple direct condition on $\mathbf{c}$, and that understanding comes only if we are willing to consider Fourier series. Let us recall some definitions.

Given a function f, defined and integrable on $(-\pi, \pi)$, the *Fourier coefficients* of f are defined by

$$\text{(5)} \qquad a_n = \frac{1}{2\pi} \int_{-\pi}^{\pi} f(\theta) e^{-in\theta} \, d\theta, \qquad -\infty < n < \infty,$$

and the series

$$\sum_{n=-\infty}^{\infty} a_n e^{in\theta}$$

is called the *Fourier series* of f. If f is square integrable, i.e. $\int |f|^2 \, d\theta < \infty$, then two things of interest happen. First $\Sigma |a_n|^2 < \infty$. This follows from *Bessel's inequality*, which says that in fact

$$\sum_{n=-\infty}^{\infty} |a_n|^2 \leq \frac{1}{2\pi} \int_{-\pi}^{\pi} |f(\theta)|^2 \, d\theta.$$

Second, the Fourier series of f converges to f in mean square, which means that the integral

$$\int_{-\pi}^{\pi} |f(\theta) - \sum_{n=-N}^{N} a_n e^{in\theta}|^2 \, d\theta$$

tends to zero as $N \longrightarrow \infty$. In this sense the Fourier series of a square-integrable function represents the function. (These facts and others we shall use can all be found in any of the standard works on the subject. See, for example, Chapter II of [4].)

As is common practice we shall denote the set of square-integrable functions by L^2. (The L is for Lebesgue, and the integrals we have been using should all be thought of as Lebesgue integrals. As usual, two functions are identified if they agree almost everywhere.) We have seen how to associate with a function $f \in L^2$ a vector $\mathbf{a} \in l^2$; $\mathbf{a}$ is the sequence of Fourier coefficients of f. Different f's give rise to different $\mathbf{a}$'s. (This follows from the mean-square convergence to f of its Fourier series.) Part of the celebrated *Riesz-Fischer theorem* (Theorem 12 of [4]) is that every $\mathbf{a} \in l^2$ arises in this way. Thus there is a one-one correspondence $f \longleftrightarrow \mathbf{a}$ between L^2 and l^2. Given f we obtain $\mathbf{a}$ by the formula (5), and given $\mathbf{a}$ we obtain f as the mean-square limit of $\sum_{-N}^{N} a_n e^{in\theta}$.

The connection between Toeplitz matrices and Fourier series is the following. Let f and g belong to L^2 and have corresponding vectors of Fourier coefficients $\mathbf{a}$ and $\mathbf{b}$, and let us find the Fourier coefficients of the product fg. The mth Fourier coefficient of fg is

$$\begin{aligned}\frac{1}{2\pi}\int f(\theta)\,g(\theta)e^{-im\theta}\,d\theta &= \frac{1}{2\pi}\int f(\theta)\left\{\sum_{n=-\infty}^{\infty} b_n e^{in\theta}\right\}e^{-im\theta}\,d\theta \\ &= \sum_{n=-\infty}^{\infty}\left\{\frac{1}{2\pi}\int f(\theta)e^{-i(m-n)\theta}\,d\theta\right\}b_n \\ &= \sum_{n=-\infty}^{\infty} a_{m-n}b_n.\end{aligned}$$

(The replacement of g by its Fourier series and the interchange of integration and summation can be justified without difficulty.) The term on the right looks very much like the left side of (4). It is about time we mentioned the word used for such an expression; it is called a *convolution*. Thus the convolution of $\mathbf{a} = (a_n)$ and $\mathbf{b} = (b_n)$ is the vector whose mth component is $\sum_{n=-\infty}^{\infty} a_{m-n}b_n$. The vector is usually denoted by $\mathbf{a} * \mathbf{b}$. The point is that under the correspondence $L^2 \longleftrightarrow l^2$ multiplication of functions corresponds to convolution of vectors. Since multiplication is surely simpler than convolution, we can expect to derive some benefit from considering Fourier series.

Let us go back now to the problem of determining when $\mathbf{T_c}$ operates on l^2. We have already seen that we must have $\Sigma\,|c_n|^2 < \infty$. Let $\phi(\theta)$ be the function with Fourier series $\Sigma\, c_n e^{in\theta}$. If $\mathbf{a} \in l^2$ with corresponding function f and $\mathbf{b} = (b_n)$ is given by (4), then as we have seen b_n are the Fourier coefficients of ϕf. The requirement that $\mathbf{b} \in l^2$ is equivalent to $\phi f \in L^2$. Thus $\mathbf{T_c}$ operates on l^2 if and only if ϕf belongs to L^2 whenever f does. It is now almost immediate that $\mathbf{T_c}$ *operates on* l^2 *if and only if the* c_n *are the Fourier coefficients of a bounded function* ϕ. Clearly if ϕ is bounded then $\phi f \in L^2$ whenever $f \in L^2$. Suppose the c_n are not the Fourier coefficients of a bounded function. Then ϕ is not essentially bounded, i.e., each of the sets

$$\{\theta\colon |\phi(\theta)| \geq k\}, \qquad k = 1, 2, \cdots$$

has positive measure. It follows that we can find an increasing sequence of positive integers $k_1, k_2, \cdots$ so that each of the sets

$$E_i = \{\theta: k_i \leq |\phi(\theta)| < k_{i+1}\}$$

has positive measure $|E_i|$. Define f by

$$f(\theta) = \begin{cases} \dfrac{1}{i|E_i|^{1/2}}, & \text{if } \theta \text{ belongs to } E_i, \\ 0, & \text{if } \theta \text{ belongs to no } E_i. \end{cases}$$

Since

$$\int |f(\theta)|^2 \, d\theta = \sum_{i=1}^{\infty} \frac{1}{i^2} < \infty$$

we have $f \in L^2$. However, on E_i

$$|\phi(\theta)f(\theta)| \geq \frac{k_i}{i|E_i|^{1/2}} \geq \frac{1}{|E_i|^{1/2}}$$

since $k_i \geq i$. Therefore

$$\int |\phi(\theta)f(\theta)|^2 \, d\theta \geq \sum_{i=1}^{\infty} 1 = \infty$$

so that $\phi f \notin L^2$ and $\mathbf{T_c}$ does not operate on L^2.

We have seen that under the correspondence $L^2 \longleftrightarrow l^2$ given by Fourier series the operation of convolution by $\mathbf{c}$ (i.e., application of the matrix $\mathbf{T_c}$) corresponds to the much simpler operation of multiplication by ϕ. Since this function plays such an important role in the study of $\mathbf{T_c}$ we shall usually write $\mathbf{T}_\phi$ instead of $\mathbf{T_c}$. Thus $\mathbf{T}_\phi$ is the Toeplitz matrix (doubly infinite in this section) corresponding to the sequence $\mathbf{c}$ of Fourier coefficients of ϕ. *We assume throughout the remainder of this article that ϕ is bounded.*

Let us return to Eq. (2). Call g the function corresponding to the given vector $\mathbf{b}$ and f the function corresponding to the sought vector $\mathbf{a}$. Then as we have seen (2) is equivalent to

$$\phi f = g \qquad \text{almost everywhere.} \tag{6}$$

The problem then is to find a necessary and sufficient condition on ϕ that for each $g \in L^2$ the equation (6) has a unique solution $f \in L^2$. Without much thought one would say that the solution is $f = g/\phi$. However, what if ϕ is zero somewhere? In fact, if ϕ is zero on a set of positive measure, (6) will not always have a solution; just take g to be a nonzero constant, for example. Given that ϕ is almost nowhere zero, $f = g/\phi$ is the only possible solution of (6). Does this belong to L^2 whenever g does? If, and only if, $1/\phi$ is essentially bounded. The proof of this is almost identical to the proof that $\mathbf{T}_\phi$ operates on l^2 if and only if ϕ is essentially bounded, and there is no point in going through it again. If $1/\phi$ is essentially bounded, then $f = g/\phi$ and $\mathbf{a}$ is obtained from $\mathbf{b}$ by convoluting with the sequence of Fourier coefficients of $1/\phi$. We have therefore established the following theorem.

THEOREM 1: *A necessary and sufficient condition that* $\mathbf{T}_\phi$ *be invertible is that* $1/\phi$ *be essentially bounded. If this holds, then* $\mathbf{T}_\phi^{-1} = \mathbf{T}_{1/\phi}$.

This theorem (for a somewhat narrower class of ϕ's) was one of the original results of Toeplitz. Actually Toeplitz's theorem dealt with the *spectrum* of $\mathbf{T}_\phi$. A complex number λ is said to belong to the spectrum of $\mathbf{T}_\phi$ if it is *not* true that the equation

$$\mathbf{T}_\phi \mathbf{a} = \lambda \mathbf{a} + \mathbf{b}$$

has a unique solution $\mathbf{a} \in l^2$ for each $\mathbf{b} \in l^2$; that is, it is not true that the matrix $\mathbf{T}_\phi - \lambda \mathbf{I}$ is invertible. Thus we have up to now been concerned with whether or not 0 belongs to the spectrum of $\mathbf{T}_\phi$. Now $\mathbf{T}_\phi - \lambda \mathbf{I}$ is the Toeplitz matrix corresponding to the sequence

$$\cdots, c_{-2}, c_{-1}, c_0 - \lambda, c_1, c_2, \cdots$$

which is the sequence of Fourier coefficients of the function $\phi(\theta) - \lambda$. From the theorem it follows that λ belongs to the spectrum of $\mathbf{T}_\phi$ if and only if $1/(\phi(\theta) - \lambda)$ is not essentially bounded. Such a λ is said to belong to the *essential range* of ϕ. (Note that if ϕ is continuous, then its essential range is its range in the usual sense.) Thus,

THEOREM 1′: *The spectrum of* $\mathbf{T}_\phi$ *is the essential range of* ϕ.

3. SEMI-INFINITE TOEPLITZ MATRICES

Before we get into the problem here, it would be helpful to make a few remarks about the doubly infinite case which should explain why the semi-infinite case is more difficult. Let $\mathbf{T}_\mathbf{c}$ and $\mathbf{T}_{\mathbf{c}'}$ be doubly infinite Toeplitz matrices, both acting on l^2, so $\mathbf{c}$ and $\mathbf{c}'$ are the sequences of Fourier coefficients of bounded functions ϕ and ϕ', respectively. Since $\mathbf{T_c}$ and $\mathbf{T_{c'}}$ correspond to multiplication by ϕ and ϕ', respectively, the product $\mathbf{T_c T_{c'}}$ corresponds to multiplication by the product $\phi\phi'$. This product has the convolution $\mathbf{c} * \mathbf{c}'$ as Fourier coefficients, so $T_{\mathbf{c}*\mathbf{c}'}$ is the Toeplitz matrix corresponding to multiplication by $\phi\phi'$. Putting these things together, one obtains the matrix identity

$$T_{\mathbf{c}}T_{\mathbf{c}'} = T_{\mathbf{c}*\mathbf{c}'}. \tag{7}$$

This is something that can be verified directly. If $\mathbf{c} = (c_n)$, $\mathbf{c}' = (c_{n'})$ then $\mathbf{T_c} = (c_{m-n})$, $\mathbf{T_{c'}} = (c'_{m-n})$ and the m, n entry of the matrix product $\mathbf{T_c T_{c'}}$ is

$$\sum_{p=-\infty}^{\infty} c_{m-p}c'_{p-n} = \sum_{p=-\infty}^{\infty} c_{m-n-p}c'_{p}, \tag{8}$$

and this is indeed the $(m - n)$th component of the convolution $\mathbf{c} * \mathbf{c}'$.

It is because of (7) that the situation is *algebraically* simple; doubly infinite Toeplitz matrices under multiplication act just like the corresponding vectors $\mathbf{c}$ under convolution. (Fourier analysis then shows that vectors under convolution act like the corresponding functions ϕ under multiplication.)

With these thoughts in mind let us turn to the semi-infinite case. We have now the matrix

$$\mathbf{T_c} = (c_{m-n})_{m,n=0}^{\infty}.$$

(We hope the reader will not be upset because we use the same notation for the doubly infinite and semi-infinite matrices.) Observe that although the subscripts m, n are nonnegative we still must have c_n defined for all n, negative as well as positive. The problem is to invert $\mathbf{T_c}$, or to solve the equation

$$\mathbf{T_c a} = \mathbf{b}$$

where $\mathbf{a}$ and $\mathbf{b}$ are vectors in l_+^2;

$$l_+^2 = \{\mathbf{a} = (a_0, a_1, \cdots) : \sum_{n=0}^{\infty} |a_n|^2 < \infty\}.$$

If we write the equation out it becomes

$$\sum_{n=0}^{\infty} c_{m-n}a_n = b_m, \qquad 0 \le m < \infty. \tag{9}$$

The difference between (9) and (4) seems slight, but the fact that all subscripts begin at 0 rather than $-\infty$ complicates matters enormously. Even the algebra is complicated. For example, identity (7) no longer holds. The reason is that if the subscript p begins at 0 in the sum on the left side of (8), the right side is no longer the $(m - n)$th component of $\mathbf{c} * \mathbf{c}'$. However, sometimes (7) is true, and we record this here for future use. Suppose $\mathbf{c}$ is such that $c_n = 0$ whenever $n > 0$. Then the m, n entry of $\mathbf{T_c T_{c'}}$ is (and now keep in mind that $m \ge 0$)

$$\sum_{p=0}^{\infty} c_{m-p}c'_{p-n} = \sum_{p=-\infty}^{\infty} c_{m-p}c'_{p-n} = \sum_{p=-\infty}^{\infty} c_{m-n-p}c'_{p},$$

which is the m, n component of $c * c'$. The reason we were able to replace 0 by $-\infty$ in the summation is that whenever $p < 0$ we have $m - p > 0$ and so $c_{m-p} = 0$. We can apply a similar argument if c' is such that $c'_n = 0$ whenever $n < 0$. Hence we can say that *we have* (7) *if either* $c_n = 0$ *whenever* $n > 0$ *or* $c'_n = 0$ *whenever* $n < 0$.

As in the doubly infinite case it must be decided when $\mathbf{T_c}$ operates on l^2_+. It so happens the answer is exactly as before: $\mathbf{T_c}$ operates on l^2_+ if and only if $\mathbf{c}$ is the sequence of Fourier coefficients of a bounded function ϕ (we won't go into the proof); and, also as before, we shall generally write $\mathbf{T}_\phi$ for $\mathbf{T}_c$. It is also true that a necessary condition for the unique solvability of (9) is that $1/\phi$ be essentially bounded. (We won't prove this either, although the proof is not very difficult.) But it is *false* that $1/\phi$ bounded implies the solvability (unique or otherwise) of (9). Let us take $\phi(\theta) = e^{i\theta}$. Then

$$c_n = \begin{cases} 0, & n \ne 1, \\ 1, & n = 1, \end{cases}$$

and the corresponding Toeplitz matrix is

$$\mathbf{S} = \begin{bmatrix} 0 & 0 & 0 & \cdots \\ 1 & 0 & 0 & \cdots \\ 0 & 1 & 0 & \cdots \\ 0 & 0 & 1 & \cdots \\ \cdots & \cdots & \cdots & \cdots \end{bmatrix}.$$

This matrix $\mathbf{S}$ is called the *shift;* it sends the vector $(a_0, a_1, \cdots)$ into the "shifted" vector $(0, a_0, a_1, \cdots)$. It is clear that the vector $\mathbf{b} = (1, 0, 0, \cdots)$, for example, is not in the range of $\mathbf{S}$, so (9) cannot be solved in this case despite the fact that $1/\phi$ is bounded. The adjoint (Hermitian transpose) $\mathbf{S}^*$ of $\mathbf{S}$ is the Toeplitz matrix

$$\begin{bmatrix} 0 & 1 & 0 & 0 & \cdots \\ 0 & 0 & 1 & 0 & \cdots \\ 0 & 0 & 0 & 1 & \cdots \\ \cdots & \cdots & \cdots & \cdots & \cdots \end{bmatrix}$$

corresponding to the function $e^{-i\theta}$. Since

$$\mathbf{S}^*(a_0, a_1, a_2, \cdots) = (a_1, a_2, a_3, \cdots)$$

it is clear that although $\mathbf{S}^*\mathbf{a} = \mathbf{b}$ always has a solution, it never has a unique solution.

The examples $\mathbf{S}$ and $\mathbf{S}^*$ show that it cannot be merely the size of ϕ that determines the invertibility of $\mathbf{T}_\phi$. In fact it is not clearly understood what does determine it. There do exist necessary and sufficient conditions on ϕ for the invertibility of $\mathbf{T}_\phi$, but these are unsatisfactory since they give us no way of deciding whether a given specific $\mathbf{T}_\phi$ is invertible or not. However, in the most interesting special cases there are satisfactory criteria, and we shall present them here since the ideas involved are of considerable interest even in themselves. Before doing this, though, we are going to present a theorem that gives at least some information in the general case.

THEOREM 2: *If $\mathbf{T}_\phi$ is invertible then $1/\phi$ is essentially bounded. If the essential range of ϕ lies entirely to one side of a line through the origin, but does not intersect the line, then $\mathbf{T}_\phi$ is invertible.*

The first part of the theorem has already been mentioned and, as we said, will not be proved here. The proof of the second part will only be described. It uses an analogue of what is known in the theory of integral equations as the *Neumann series.* The idea is this. Suppose we wish to invert $\mathbf{I} - \mathbf{T}$. In our case $\mathbf{I}$ is the identity matrix and $\mathbf{T}$ is a Toeplitz matrix, but let's forget about this. If $\mathbf{I}$ were the number 1 and $\mathbf{T}$ were a number of absolute value less than 1, then the inverse of $\mathbf{I} - \mathbf{T}$ would be given by the convergent geometric series

$$\mathbf{I} + \mathbf{T} + \mathbf{T}^2 + \cdots.$$

Now it turns out that if $\mathbf{T}$ is "sufficiently small" in an appropriate sense then this series will converge, also in an appropriate sense, and its sum will be $(\mathbf{I} - \mathbf{T})^{-1}$. This is true for quite general topological-algebraic objects of which infinite matrices such as ours are special cases. A Toeplitz matrix will be "sufficiently small" if $||\phi||_\infty < 1$. Here $||\phi||_\infty$, the essential supremum of $|\phi|$, is the smallest number M for which the inequality $|\phi(\theta)| \leq M$ holds almost everywhere. Thus $\mathbf{T}_\phi - \mathbf{I}$ will be invertible if $||\phi||_\infty < 1$. Now suppose we have a line L through the origin so that the essential range of ϕ lies entirely to one side of L. Let α be the unique number of absolute value 1 which lies on this side of L and such that the line joining α to 0 is perpendicular to L. Since the essential range of ϕ is a bounded closed set it is easy to see geometrically (and is even true) that for sufficiently small positive ϵ we shall have

$$||\alpha - \epsilon\phi||_\infty < 1.$$

This is equivalent to

$$||1 - \frac{\epsilon}{\alpha}\phi||_\infty < 1$$

and so it implies that the matrix

$$\mathbf{I} - \mathbf{T}_{1-(\epsilon/\alpha)\phi} = \frac{\epsilon}{\alpha}\mathbf{T}_\phi$$

is invertible. Therefore $\mathbf{T}_\phi$ is invertible.

There is a statement concerning the spectrum of T_ϕ that can be derived from Theorem 2. It is most simply and elegantly stated in terms of convex sets. A set is *convex* if it contains along with any pair of points the entire line segment joining these points. The *convex hull* of a set is the smallest convex set containing the given set. Now the convex hull of an arbitrary set in the plane is the intersection of all half-planes containing it. This fact and Theorem 2 give the following.

THEOREM 2′: *The spectrum of* $\mathbf{T}_\phi$ *contains the essential range of* ϕ *and is contained in the convex hull of the essential range of* ϕ.

Both parts of Theorem 2′ are the best possible in the following sense. Let S be an arbitrary bounded closed set in the plane. Then there is a ϕ whose essential range is S and such that the spectrum of $\mathbf{T}_\phi$ is the full convex hull of S. But given $\alpha \notin S$ there is a ϕ whose essential range is S and such that α does not belong to the spectrum of $\mathbf{T}_\phi$. This last statement, of course, does not imply that there is a ϕ whose corresponding $\mathbf{T}_\phi$ has spectrum exactly S. In fact it is known that the spectrum of any semi-infinite Toeplitz matrix is a connected set, so if S is disconnected any such spectrum which contains S must contain S properly.

We now pass to the special cases in which the spectrum can be determined exactly.

(a) *The Triangular Case.* A triangular matrix is one which has only zeros above, or below, the main diagonal. The statement that only zeros appear above the main diagonal of $\mathbf{T}_\mathbf{c}$ is equivalent to $c_n = 0$ for $n < 0$. In this case the corresponding function ϕ has as Fourier series

$$\sum_{n=0}^{\infty} c_n e^{in\theta}$$

and is of what we shall call *analytic type*, since it has associated with it an analytic function

$$\Phi(z) = \sum_{n=0}^{\infty} c_n z^n, \qquad |z| < 1.$$

Similarly the statement that only zeros appear below the main diagonal of $\mathbf{T}_\mathbf{c}$ is equivalent to $c_n = 0$ for $n > 0$ and we shall say that ϕ is of *coanalytic type.* Now recall the little result we established above which extended (7) to the semi-infinite case under certain circumstances. In terms of functions it says that *if either ϕ is of coanalytic type or ϕ' is of analytic type then* $\mathbf{T}_\phi \mathbf{T}_{\phi'} = \mathbf{T}_{\phi\phi'}$.

Consider a function ϕ of analytic type. It goes without saying that ϕ is assumed bounded. Let us also assume that $1/\phi$ is bounded since this is certainly necessary for invertibility of T_ϕ. The natural try for $\mathbf{T}_\phi^{-1}$ would be $\mathbf{T}_{1/\phi}$. Indeed, since ϕ is of analytic type,

$$\mathbf{T}_{1/\phi} \mathbf{T}_\phi = \mathbf{T}_1 = \mathbf{I},$$

so that $\mathbf{T}_{1/\phi}$ is a left inverse for $\mathbf{T}_\phi$. In general we cannot say more. (Consider the case of $\phi(\theta) = e^{i\theta}$, which is of analytic type.) We would like to say that also

$$\mathbf{T}_\phi \mathbf{T}_{1/\phi} = \mathbf{T}_1 = \mathbf{I},$$

but since ϕ is not of coanalytic type we cannot say it, *unless, of course,* $1/\phi$ *is of analytic type.* We have just proved one-half of the following theorem.

THEOREM 3: *Assume* ϕ *is of analytic type. Then* $\mathbf{T}_\phi$ *is invertible if and only if* $1/\phi$ *is bounded and of analytic type. In this case* $\mathbf{T}_\phi^{-1} = \mathbf{T}_{1/\phi}$.

What is left for us to prove is that if $\mathbf{T}_\phi$ is invertible then $1/\phi$ is of analytic type. Under the assumption that $\mathbf{T}_\phi$ is invertible we can solve the equation

$$\sum_{n=0}^{\infty} c_{m-n} a_n = \begin{cases} 1, & m = 0, \\ 0, & m > 0, \end{cases} \tag{10}$$

for some vector $\mathbf{a} = (a_0, a_1, \cdots) \in l^2_+$. Because ϕ is of analytic type, $c_{m-n} = 0$ whenever $n \geq 0$ and $m < 0$. It follows that (10) can be strengthened to read

$$\sum_{n=0}^{\infty} c_{m-n} a_n = \begin{cases} 1, & m = 0, \\ 0, & m > 0 \text{ or } m < 0. \end{cases}$$

Next define a_n to be zero if $n < 0$. Then

$$\sum_{n=-\infty}^{\infty} c_{m-n} a_n = \begin{cases} 1, & m = 0, \\ 0, & m \neq 0. \end{cases}$$

Since convolution of doubly infinite vectors corresponds to multiplication of functions, it follows that the function 1 is the product of ϕ and the function f whose Fourier series is

$$\sum_{n=-\infty}^{\infty} a_n e^{in\theta} = \sum_{n=0}^{\infty} a_n e^{in\theta}.$$

Thus $1/\phi = f$ is of analytic type.

The statement of Theorem 3 naturally raises the question of how to determine when $1/\phi$ is of analytic type. Of course this means

$$\int \frac{1}{\phi(\theta)} e^{-in\theta} \, d\theta = 0, \qquad n = -1, -2, \cdots,$$

but this may be difficult to verify in practice. It turns out that there is a fairly simple criterion in terms of the analytic function Φ. *Let ϕ be of analytic type with associated analytic function $\Phi(z)$, $|z| < 1$. Then $1/\phi$ is essentially bounded and of analytic type if and only if $1/\Phi(z)$ is bounded in $|z| < 1$.* In particular Φ can have no zeros inside the unit circle. Thus if $\phi(\theta) = e^{i\theta}$, then $\Phi(z) = z$ and $1/\phi(\theta) = e^{-i\theta}$ is not of analytic type. To prove the assertion we are going to need a couple of facts from the theory of Abel summability of Fourier series.

Suppose we have a function $f \in L^2$ with associated Fourier series $\Sigma\, a_n e^{in\theta}$. Given a number r in $0 < r < 1$ the function

$$f_r(\theta) = \sum_{n=-\infty}^{\infty} a_n r^{|n|} e^{in\theta}$$

is called the rth Abel mean of the Fourier series of f. Note that, since $\Sigma\, |a_n|^2 < \infty$, an application of Schwarz's inequality shows that the series defining $f_r(\theta)$ converges for all θ. The facts about $f_r(\theta)$ we shall need are these:

(i) If $|f(\theta)| \leq M$ for almost every θ then $|f_r(\theta)| \leq M$ for all θ and r.

(ii) For almost every θ we have

$$\lim_{r \to 1-} f_r(\theta) = f(\theta).$$

The first of these facts follows from the representation of $f_r(\theta)$ as a Poisson integral involving $f(\theta)$; the second is the analogue for Abel summability of the Fejér-Lebesgue theorem. (We refer the interested reader to Sec. 5.9 of [4] for details.)

Let us take the special case of our function ϕ of analytic type and having Fourier series $\sum_{n=0}^{\infty} c_n e^{in\theta}$. The corresponding Abel means are

$$\phi_r(\theta) = \sum_{n=0}^{\infty} c_n r^n e^{in\theta} = \Phi(re^{i\theta}).$$

From fact (ii),

$$\lim_{r \to 1-} \Phi(re^{i\theta}) = \phi(\theta) \qquad \text{almost everywhere.}$$

Now suppose that

$$\frac{1}{|\Phi(z)|} \le M, \qquad |z| < 1.$$

Then for almost every θ

$$\frac{1}{|\phi(\theta)|} = \lim_{r \to 1-} \frac{1}{|\Phi(re^{i\theta})|} \le M,$$

so that $1/\phi$ is essentially bounded. Moreover, for $n = -1, -2, \cdots$ we have

$$\int \frac{1}{\phi(\theta)} e^{-in\theta}\, d\theta = \lim_{r \to 1-} \int \frac{1}{\Phi(re^{i\theta})} e^{-in\theta}\, d\theta = 0,$$

since for each r the function $1/\Phi(re^{i\theta})$ is of analytic type. The passage to the limit under the integral sign is justified by the Lebesgue bounded convergence theorem.

We have shown that $1/\Phi$ bounded in $|z| < 1$ implies $1/\phi$ bounded and of analytic type. Conversely assume $1/\phi$ is bounded and of analytic type, and let it have Fourier series $\sum_{n=0}^{\infty} \gamma_n e^{in\theta}$. Since ϕ has Fourier series $\sum_{n=0}^{\infty} c_n e^{in\theta}$ and since $\phi \cdot (1/\phi) = 1$, it follows that

$$\sum_{n=0}^{m} c_{m-n} \gamma_n = \begin{cases} 1, & m = 0, \\ 0, & m \neq 0. \end{cases} \tag{11}$$

(The sum is taken up to m rather than ∞, since $c_{m-n} = 0$ whenever $n > m$.) This implies that the analytic function

$$\Psi(z) = \sum_{n=0}^{\infty} \gamma_n z^n, \qquad |z| < 1,$$

corresponding to $1/\phi$ is exactly $1/\Phi(z)$. One need only multiply the series for Φ and Ψ and use (11). That $1/\Phi = \Psi$ is bounded in $|z| < 1$ then follows from (i) and the boundedness of $1/\phi$.

One can now restate Theorem 3 as follows.

THEOREM 3′: *Assume ϕ is of analytic type. Then $\mathbf{T}_\phi$ is invertible if and only if the corresponding analytic function Φ is bounded away from zero inside the unit circle.*

A statement equivalent to Theorem 3′ is

THEOREM 3″: *Assume ϕ is of analytic type. Then the spectrum of $\mathbf{T}_\phi$ is the closure of the range of $\Phi(z)$ for $|z| < 1$.*

We should point out the following interesting special case. If ϕ is continuous and $\phi(-\pi) = \phi(\pi)$, then Φ can be extended so as to be continuous on the *closed* disc $|z| \leq 1$. In fact we will then have $\Phi(e^{i\theta}) = \phi(\theta)$. In this case the spectrum of $\mathbf{T}_\phi$ is exactly the range of $\Phi(z)$ for $|z| \leq 1$. An example of a ϕ that is bounded and of analytic type but nevertheless discontinuous may be hard to think of, except for one obtained trivially by taking a continuous one and modifying it on a set of measure zero. Perhaps the simplest nontrivial example is

$$\phi(\theta) = \exp\left\{-i \cot \frac{\theta}{2}\right\}.$$

This is discontinuous at $\theta = 0$. (Of course it is undefined at $\theta = 0$, but no value assigned to ϕ at 0 would make it continuous there.) It would be instructive for the reader to prove that this function, which is clearly bounded, is of analytic type.

We have considered here only triangular Toeplitz matrices arising from functions of analytic type. What if ϕ is of coanalytic type? It happens that the invertibility of $\mathbf{T}_\phi$ is equivalent to that of $\mathbf{T}_{\bar\phi}$, where $\bar\phi$ is the complex conjugate of ϕ. The simplest way to see this is to observe that since $\bar\phi$ has nth Fourier coefficient $\bar c_{-n}$ (by direct computation), one has $\mathbf{T}_{\bar\phi} = \mathbf{T}_\phi^*$, the adjoint of $\mathbf{T}_\phi$. Thus if $\mathbf{T}_\phi$ is invertible we deduce from

$$\mathbf{T}_\phi^{-1}\mathbf{T}_\phi = \mathbf{T}_\phi\mathbf{T}_\phi^{-1} = \mathbf{I},$$

upon taking adjoints, that

$$\mathbf{T}_{\bar\phi}(\mathbf{T}_\phi^{-1})^* = (\mathbf{T}_\phi^{-1})^*\mathbf{T}_{\bar\phi} = \mathbf{I};$$

that is to say, $(\mathbf{T}_\phi^{-1})^*$ is the inverse of $\mathbf{T}_{\bar\phi}$. Therefore if ϕ is of coanalytic type the invertibility of $\mathbf{T}_\phi$ can be decided by applying Theorem 3 or 3′ to the function $\bar\phi$ of analytic type.

(b) *The Hermitian Case.* A matrix is Hermitian if it is equal to its adjoint. Since $\mathbf{T}_\phi^* = \mathbf{T}_{\bar\phi}$, this is equivalent to ϕ being real-valued. Let m and M be respectively the largest and smallest numbers for which

$$m \leq \phi(\theta) \leq M \qquad \text{almost everywhere.}$$

THEOREM 4: *The spectrum of* $\mathbf{T}_\phi$ *is the closed interval* $[m, M]$.

That the spectrum of $\mathbf{T}_\phi$ is contained in $[m, M]$ follows from Theorem 2, since $[m, M]$ is a convex set containing the essential range of ϕ. Let us see why the spectrum is the full interval. Since m and M belong to the essential range of ϕ, Theorem 2 tells us that they belong to the spectrum. In the doubly infinite case this is all we could say in general, because if ϕ is discontinuous the range of ϕ may very well consist of only the numbers m and M. (See Theorem 1.) However, here it is true that for no number λ in $m < \lambda < M$ is $\mathbf{T}_\phi - \lambda\mathbf{I}$ invertible. For suppose otherwise, and assume (as we may without loss of generality) that $\lambda = 0$. Then we have

$$m < 0 < M \tag{12}$$

but nevertheless $\mathbf{T}_\phi$ is invertible. By invertibility we can find a vector $\mathbf{a} \in l^2$ so that

$$\sum_{n=0}^{\infty} c_{m-n}a_n = \begin{cases} 1, & m = 0, \\ 0, & m > 0. \end{cases} \tag{13}$$

Then if f is the function with Fourier series $\sum_{n=0}^{\infty} a_n e^{in\theta}$, f is of analytic type and ϕf is of coanalytic type. (Identity (13) says that the mth Fourier coefficient of ϕf vanishes if $m > 0$.) Therefore $\phi|f|^2 = \phi f\bar{f}$ is also of coanalytic type. But the only real-valued function of coanalytic type is a constant. Therefore

$$\phi(\theta)|f(\theta)|^2 = c \qquad \text{almost everywhere.}$$

Now $c \neq 0$. For if $c = 0$ we would have $\phi f = 0$ almost everywhere, and this is not so since the 0th Fourier coefficient of ϕf is 1. Therefore either $c > 0$ or $c < 0$. But since $|f|^2$ is nonnegative, $c > 0$ implies that $\phi \geq 0$ almost everywhere. This means $m \geq 0$ and contradicts (12). Similarly $c < 0$ leads to a contradiction.

(c) *The Continuous Case.* In this section when we say "ϕ is continuous" we shall really mean "ϕ is continuous and $\phi(-\pi) = \phi(\pi)$." This extra condition insures that the function on the unit circle $|z| = 1$ which takes the value $\phi(\theta)$ at $e^{i\theta}$ is continuous; and this is what is important. To prepare us for the general situation, we shall take a special case where we know the answer.

Suppose $\phi(\theta) = \sum_{n=0}^{\infty} c_n e^{in\theta}$, where the c_n tend rapidly to zero as $n \longrightarrow \infty$; in fact we assume that for some $R > 1$

$$|c_n| < R^{-n}$$

for all sufficiently large n. Then ϕ is both continuous and of analytic type. Moreover, the corresponding analytic function Φ is even analytic in the region $|z| < R$, which contains the unit circle. Now by Theorem 3′, $\mathbf{T}_\phi$ is invertible if and only if Φ is bounded away from zero in $|z| < 1$, which is equivalent to Φ being nonzero in $|z| \leq 1$. Of course, saying Φ has no zeros on the boundary $|z| = 1$ is exactly the same as saying ϕ is never zero, and this we take for granted. Now given a function $\Phi(z)$ analytic inside and on a simple closed curve C (in our case C is $|z| = 1$) there is a way of determining the number of zeros N of Φ inside C from the values of Φ on C. In fact

$$2\pi N = \underset{C}{\Delta} \arg \Phi(z),$$

the variation of the argument of $\Phi(z)$ around C. (See, for example, Sec. 3.41 of [10].) We can therefore say that in our case $\mathbf{T}_\phi$ is invertible if and only if

$$\underset{-\pi \leq \theta \leq \pi}{\Delta} \arg \phi(\theta) = 0.$$

The number

$$\frac{1}{2\pi} \underset{-\pi \leq \theta \leq \pi}{\Delta} \phi(\theta),$$

which we shall denote by $I(\phi)$, is known as the *index* of ϕ or the *winding number* of ϕ. It represents the number of times the curve traced out by ϕ in the complex plane (the curve is closed since $\phi(-\pi) = \phi(\pi)$) winds around 0.

THEOREM 5: *Assume ϕ is continuous. Then $\mathbf{T}_\phi$ is invertible if and only if ϕ is nowhere zero and $I(\phi) = 0$.*

Before giving the proof we must caution the reader that $I(\phi) = 0$ does not mean that the range of ϕ does not surround 0. If we set

$$\phi(\theta) = \begin{cases} e^{-2i\theta} & -\pi \leq \theta \leq 0, \\ e^{2i\theta}, & 0 \leq \theta \leq \pi, \end{cases}$$

then the range of ϕ is the entire unit circle and yet $I(\phi) = 0$.

In order to present the main idea of the proof we shall first consider a fairly special case. We shall suppose ϕ is the exponential of a trigonometric polynomial,

$$\phi(\theta) = e^{\sum_{n=-N}^{N} \alpha_n e^{in\theta}}.$$

Of course $\phi(\theta) \neq 0$. Moreover

$$\arg \phi(\theta) = \mathscr{I} \sum_{n=-N}^{N} \alpha_n e^{in\theta},$$

which is a continuous function of period 2π, so $I(\phi) = 0$. We shall show that all these $\mathbf{T}_\phi$ are invertible. This will be done by factoring ϕ into two functions, respectively of analytic and coanalytic type, each of which gives rise to an invertible Toeplitz matrix. In fact let us define

$$\phi_+(\theta) = e^{\sum_{n=0}^{N} \alpha_n e^{in\theta}}, \qquad \phi_-(\theta) = e^{\sum_{n=-N}^{-1} \alpha_n e^{in\theta}}.$$

Clearly $\phi(\theta) = \phi_-(\theta)\phi_+(\theta)$. The function

$$\Phi_+(z) = e^{\sum_{n=0}^{N} \alpha_n z^n}$$

is analytic throughout the plane. Therefore by the Cauchy integral theorem we have for each $n = -1, -2, \cdots$

$$0 = \frac{1}{2\pi i} \int_{|z|=1} \Phi_+(z) z^{-n-1}\, dz = \frac{1}{2\pi} \int_{-\pi}^{\pi} \phi_+(\theta) e^{-in\theta}\, d\theta,$$

and ϕ_+ is of analytic type. Similarly $\bar{\phi}_-$ is of analytic type so ϕ_- is of coanalytic type. Since $\Phi_+(z)$ is nowhere zero, it follows from Theorem 3′ that $\mathbf{T}_{\phi_+}$ is invertible, and similarly so is $\mathbf{T}_{\phi_-}$. But then $\mathbf{T}_\phi = \mathbf{T}_{\phi_-}\mathbf{T}_{\phi_+}$ is also invertible. Notice that since Theorem 3 gives the inverses of $\mathbf{T}_{\phi_+}$ and $\mathbf{T}_{\phi_-}$ we can write down the inverse of $\mathbf{T}_\phi$:

$$\mathbf{T}_\phi^{-1} = \mathbf{T}_{1/\phi_+}\mathbf{T}_{1/\phi_-}. \tag{14}$$

It is tempting to go further and say $\mathbf{T}_\phi^{-1} = \mathbf{T}_{1/\phi}$. This we cannot do, since in general neither is $1/\phi_+$ of coanalytic type nor is $1/\phi_-$ of analytic type. In fact the identity $\mathbf{T}_\phi^{-1} = \mathbf{T}_{1/\phi}$ holds only if ϕ is either of analytic or coanalytic type.

The general case can be handled with only a little more difficulty than the special case. Suppose that ϕ is an arbitrary continuous function which is nowhere zero and satisfies $I(\phi) = 0$. Then it has a continuous logarithm which we shall denote by $\log \phi(\theta)$. (Here is where $I(\phi) = 0$ comes in; otherwise we would not have $\log \phi(-\pi) = \log \phi(\pi)$.) We need next a theorem which is an analogue for trigonometric polynomials of the famous Weierstrass approximation theorem. *Given a continuous function* $\psi(\theta)$ *and an* $\epsilon > 0$ *there is a trigonometric polynomial* $p(\theta)$ *such that*

$$|p(\theta) - \psi(\theta)| \leq \epsilon, \qquad -\pi \leq \theta \leq \pi.$$

(For a proof of this see Theorem 24 of [4].) In our case $\psi(\theta) = \log \phi(\theta)$ and we can afford to be fairly generous with ϵ. Let us find a trigonometric polynomial

$$p(\theta) = \sum_{n=-N}^{N} \alpha_n e^{in\theta}$$

so that

$$(15) \qquad |p(\theta) - \log \phi(\theta)| \leq \frac{\pi}{4}, \qquad -\pi \leq \theta \leq \pi.$$

We proceed now very much as we did before. If we set

$$\phi_+(\theta) = e^{\sum_{n=0}^{N} \alpha_n e^{in\theta}}, \qquad \phi_-(\theta) = e^{\sum_{n=-N}^{-1} \alpha_n e^{in\theta}},$$

then we have $\phi_-\phi_+ = e^p$, so $\phi = \phi_-(\phi e^{-p})\phi_+$. Since ϕ_- is of coanalytic type *and* ϕ_+ is of analytic type, we obtain

$$\mathbf{T}_\phi = \mathbf{T}_{\phi_-}\mathbf{T}_{(\phi e^{-p})\phi_+} = \mathbf{T}_{\phi_-}\mathbf{T}_{\phi e^{-p}}\mathbf{T}_{\phi_+}.$$

As before, $\mathbf{T}_{\phi_-}$ and $\mathbf{T}_{\phi_+}$ are invertible. Inequality (15) implies that the range of ϕe^{-p} lies to the right of the imaginary axis, so $\mathbf{T}_{\phi e^{-p}}$ is invertible by Theorem 2. Therefore $\mathbf{T}_\phi$ is invertible. Moreover, if this is of any interest,

$$\mathbf{T}_\phi^{-1} = \mathbf{T}_{1/\phi_+}\mathbf{T}^{-1}_{\phi e^{-\nu}}\mathbf{T}_{1/\phi_-},$$

which is a generalization of (14).

We have so far shown that $I(\phi) = 0$ is sufficient for the invertibility of T_ϕ. To show that the condition is also necessary, suppose $I(\phi) = n \neq 0$, and consider first the case $n > 0$. We have then

$$\phi(\theta) = \phi(\theta)e^{-in\theta}\cdot e^{in\theta}.$$

Since $e^{in\theta}$ is of analytic type,

$$\mathbf{T}_\phi = \mathbf{T}_{\phi e^{-in\theta}}\mathbf{T}_{e^{in\theta}} = \mathbf{T}_{\phi e^{-in\theta}}\mathbf{S}^n, \tag{16}$$

where $\mathbf{S}^n$ is the nth power of the shift. Since $I(\phi e^{-in\theta}) = I(\phi) - n = 0$, we know that $\mathbf{T}_{\phi e^{-in\theta}}$ is invertible. It follows from this and (16) that if $\mathbf{T}_\phi$ were invertible then $\mathbf{S}^n$ would be. Since $\mathbf{S}^n$ is not, neither is $\mathbf{T}_\phi$. A similar argument which uses powers of $\mathbf{S}^*$ takes care of the case $n < 0$ and completes the proof of Theorem 5.

What of the spectrum of a Toeplitz matrix with continuous ϕ? Well, λ belongs to the spectrum of $\mathbf{T}_\phi$ if and only if $\mathbf{T}_{\phi-\lambda}$ is not invertible, and this happens if either $\phi(\theta) - \lambda$ is zero somewhere or $I(\phi(\theta) - \lambda) \neq 0$. This can be interpreted (somewhat loosely) as follows: *The spectrum of* $\mathbf{T}_\phi$ *consists of the curve traced out by* ϕ *together with the points the curve winds around.*

There is an extension of this to certain discontinuous ϕ's which we would like to mention without proof. Suppose you had a pencil and were required to draw the curve traced out by $\phi(\theta)$ as θ traversed the interval $-\pi \leq \theta \leq \pi$, but you were not allowed to lift your pencil in the process. If ϕ were continuous there would be no difficulty, at least in principle. But suppose ϕ were discontinuous and had a discontinuity of jump type at θ_0, say. Then to get from $\phi(\theta_0 -)$ to $\phi(\theta_0 +)$ you would have to add a piece to the curve that did not really belong there, and to get this over with as quickly as possible you would add the smallest possible piece, namely the line segment joining $\phi(\theta_0 -)$ and $\phi(\theta_0 +)$. Now if ϕ had finitely many jump discontinuities and you performed this task adroitly you would have drawn a closed curve consisting of the range of ϕ and some line segments. *The spectrum of* $\mathbf{T}_\phi$ *consists of this curve together with the points the curve winds around.*

This determines the spectrum of a very large class of Toeplitz matrices, one containing (one might argue) any Toeplitz matrix that one might reasonably expect to encounter. However it must be pointed out that there are a lot of functions, and almost none of them are so nice as to be continuous except for jump discontinuities. In fact the general problem of invertibility of semi-infinite Toeplitz matrices remains unresolved.

4. FINITE TOEPLITZ MATRICES

In this section we shall consider $N \times N$ Toeplitz matrices

$$\begin{bmatrix} c_0 & c_{-1} & \cdots & c_{-N+1} \\ c_1 & c_0 & \cdots & c_{-N+2} \\ \cdots & \cdots & \cdots & \cdots \\ c_{N-1} & c_{N-2} & \cdots & c_0 \end{bmatrix}. \tag{17}$$

It is very easy to determine whether a given finite matrix is invertible. Invertibility is equivalent to the nonvanishing of the determinant. More generally, the spectrum of a finite matrix is the set of its eigenvalues, which are the roots of its characteristic equation. Even if the matrix is Toeplitz, this situation cannot be simplified. Therefore, what we shall be concerned with here is not the determination of the spectrum of the matrix (17), but rather the behavior of this spectrum as the size of the matrix increases. As before, we shall assume that $c_n(-\infty < n < \infty)$ are the Fourier coefficients of a bounded function ϕ. To indicate its dependence on both ϕ and N we shall denote the matrix (17) by $\mathbf{T}_{\phi,N}$.

The first question one might ask is, "What is the set 'filled in' by the spectrum of $\mathbf{T}_{\phi,N}$ as $N \longrightarrow \infty$?" There is very little known about this for general ϕ. However, in this section we shall be concerned exclusively with real ϕ (Hermitian $\mathbf{T}_\phi$) and the answer in this case is, "The spectrum of the corresponding semi-infinite Toeplitz matrix." Thus if, as before, m and M are respectively the largest and smallest numbers for which

$$m \leq \phi(\theta) \leq M \qquad \text{almost everywhere,}$$

then the spectrum of $\mathbf{T}_{\phi,N}$ fills in the interval $[m, M]$. More precisely, if we denote the eigenvalues of $\mathbf{T}_{\phi,N}$ by $\lambda_i (i = 1, \cdots, N)$, then each λ_i satisfies

$$m \leq \lambda_i \leq M, \tag{18}$$

and any subinterval of $[m, M]$ will contain some λ_i if N is sufficiently large. (Strictly speaking, λ should have two subscripts, N which tells us we are dealing with the matrix $\mathbf{T}_{\phi,N}$, and i which tells us we are dealing with the ith eigenvalue of this matrix. We have suppressed the subscript N and hope this will cause no confusion. To add another parenthetical remark here, the list $\lambda_1, \cdots, \lambda_N$ may contain repetitions; an eigenvalue of multiplicity r appears r times.)

One half of this is quite easy. If we observe that Theorem 2′ holds in the finite case then (18) is immediate. The reason Theorem 2′ does hold here is that its proof made no use whatever of the size of $\mathbf{T}_\phi$; in fact the theorem holds for any matrix that is a section, symmetric about the main diagonal, of a doubly infinite Toeplitz matrix. We do not propose to prove here the other half, that all of $[m, M]$ is filled in. Rather we are going to present a theorem due to G. Szegö which tells us in a sense how the interval is filled in. Szegö's theorem says, roughly, that the numbers $\lambda_1, \cdots, \lambda_N$ are asymptotically distributed just like the values of ϕ at N equally spaced points in $(-\pi, \pi)$; thus $\lambda_1, \cdots, \lambda_N$ fill in $[m, M]$ in the same way that the numbers

$$\phi\left(\frac{2i - N}{N}\pi\right), \qquad i = 1, \cdots, N \tag{19}$$

do. Here is the precise statement.

THEOREM 6: *Let F be any continuous function defined on $[m, M]$. Then*

$$\lim_{N\to\infty} \frac{F(\lambda_1) + \cdots + F(\lambda_N)}{N} = \frac{1}{2\pi}\int_{-\pi}^{\pi} F[\phi(\theta)]\, d\theta.$$

This doesn't look at all like what we just said, so before we prove the theorem we ought to try to justify our interpretation of it. Suppose that λ_i were exactly the quantities (19). Then we would have

$$\frac{2\pi}{N}\{F(\lambda_1)+\cdots+F(\lambda_N)\} = \sum_{i=1}^{N} F\left[\phi\left(\frac{2i-N}{N}\pi\right)\right]\frac{2\pi}{N}. \tag{20}$$

The right side is just a Riemann sum for the integral in the statement of the theorem, and (at least if ϕ is continuous) the conclusion of the theorem would hold. Of course the λ_i are not the quantities (19), but the theorem says they are sufficiently like them that the sum on the left of (20) behaves like a Riemann sum for the integral no matter what F is.

To prove the theorem we shall do what is often done in analysis: consider first a large number of special cases (in some sense a dense set of special cases) which can be handled by special methods, and then extend the result to the general case by an approximation argument. Our special cases are the functions $F(x) = x^s$ ($s = 0, 1, 2, \cdots$). Thus what is to be proved first is

$$\lim_{N\to\infty}\frac{1}{N}\sum_{i=1}^{N}\lambda_i^s = \frac{1}{2\pi}\int[\phi(\theta)]^s\,d\theta. \tag{21}$$

The case $s = 0$ is completely trivial since (21) then just says $1 = 1$. Let us consider next $s = 1$. The quantity $\Sigma\,\lambda_i$ is, by a well-known theorem from the theory of matrices, the *trace* of $\mathbf{T}_{\phi,N}$, the sum of its diagonal entries. Since each diagonal entry is c_0 and since there are N of them, we have $\Sigma\,\lambda_i = Nc_0$ and the left side of (21) is c_0 no matter what N is. But the same is true of the right side of (21), since c_0 is the 0th Fourier coefficient of ϕ. Therefore (21) holds in this case also.

To avoid getting involved with too many subscripts we shall not consider general s but rather a typical one, $s = 3$. First let us compute the right side of (21) in terms of quantities c_n. Since $[\phi(\theta)]^3$ has as Fourier coefficients the threefold convolution of $\{c_n\}$ with itself, the nth Fourier coefficient of $[\phi(\theta)]^3$ is

$$\sum_{p,q=-\infty}^{\infty} c_{n-p}c_{p-q}c_q.$$

Since the right side of (21) is the 0th Fourier coefficient of $[\phi(\theta)]^3$,

$$\frac{1}{2\pi}\int[\phi(\theta)]^3\,d\theta = \sum_{p,q=-\infty}^{\infty} c_{-p}c_{p-q}c_q. \tag{22}$$

Next, since $\Sigma\, \lambda_i^3$ is the trace of $\mathbf{T}^3_{\phi,N}$, which is the sum of its diagonal entries, let us write down its n, n entry. It is

$$\sum_{p,q=1}^{N} c_{n-p}c_{p-q}c_{q-n},$$

which, after replacing $p - n$, $q - n$ by p, q, respectively, becomes

$$\sum_{p,q=1-n}^{N-n} c_{-p}c_{p-q}c_{q}.$$

This looks very much like the right side of (22). In fact, if N and n are very large (but N much larger than n) this will be very close to the right side of (22). Loosely speaking, if N is very large then almost all diagonal entries of $\mathbf{T}^3_{\phi,N}$ are very close to the right side of (22). This can be made quite precise. It follows that the average of the diagonal entries, which is just $N^{-1}\,\Sigma\,\lambda_i^3$, is also very close to the right side of (22). That is,

$$\lim_{N\to\infty} \frac{1}{N} \sum_{i=1}^{N} \lambda_i^3 = \sum_{p,q=-\infty}^{\infty} c_{-p}c_{p-q}c_q = \frac{1}{2\pi}\int [\phi(\theta)]^3\, d\theta.$$

It follows immediately from (21) that the theorem holds for any polynomial F. The passage to general continuous F will depend on the Weierstrass approximation theorem. Suppose we have a polynomial P such that

$$P(x) \le F(x), \qquad m \le x \le M.$$

Then, since each λ_i lies in $[m, M]$, we have

$$\frac{1}{N}\,\Sigma\, P(\lambda_i) \le \frac{1}{N}\,\Sigma\, F(\lambda_i).$$

Since the theorem is known to be true for P,

$$\lim_{N\to\infty} \frac{1}{N}\,\Sigma\, P(\lambda_i) = \frac{1}{2\pi}\int P[\phi(\theta)]\, d\theta.$$

Therefore we can say

$$\liminf_{N\to\infty} \frac{1}{N}\,\Sigma\, F(\lambda_i) \ge \frac{1}{2\pi}\int P[\phi(\theta)]\, d\theta. \tag{23}$$

This holds for any polynomial P which is always less than or equal to F. Now there is a P which satisfies

$$(24) \qquad F(x) - \epsilon \leq P(x) \leq F(x), \qquad m \leq x \leq M.$$

(One need only find a P which satisfies

$$\left| P(x) - \left(F(x) - \frac{\epsilon}{2} \right) \right| \leq \frac{\epsilon}{2},$$

and one can do this, by the Weierstrass approximation theorem.) It follows from (23) and (24) that

$$\liminf_{N \to \infty} \frac{1}{N} \Sigma\, F(\lambda_i) \geq \frac{1}{2\pi} \int F[\phi(\theta)]\, d\theta - \epsilon.$$

Since ϵ is arbitrary, it may be removed without destroying the inequality. Similarly we can show

$$\limsup_{N \to \infty} \frac{1}{N} \Sigma\, F(\lambda_i) \leq \frac{1}{2\pi} \int F[\phi(\theta)]\, d\theta,$$

and the two inequalities combine to give the conclusion of the theorem.

There are a couple of corollaries of the theorem which are of some interest. Suppose we have an interval $[a, b]$ contained in $[m, M]$ and ask for the approximate number of eigenvalues of $\mathbf{T}_{\phi,N}$ which belong to it. If the eigenvalues were equally spaced throughout $[m, M]$, then we would expect the proportion of eigenvalues which lie in $[a, b]$ to be approximately equal to the ratio $(b - a)/(M - m)$. But the eigenvalues are not equally spaced. In fact they are distributed like the values of ϕ at N equally spaced points, and this suggests that the proportion of eigenvalues lying in $[a, b]$ is (in the limit) the proportion of θ's for which $\phi(\theta)$ lies in $[a, b]$. This is true, and is the content of the first corollary. Let us introduce some notation. Denote by $\nu(a, b, N)$ the number of eigenvalues of $\mathbf{T}_{\phi,N}$ which lie in $[a, b]$, and by $\Omega(a, b)$ the measure of the set of θ's for which $\phi(\theta)$ lies in $[a, b]$.

COROLLARY 1: *Assume*

$$\{\theta \colon \phi(\theta) = a \quad or \quad \phi(\theta) = b\}$$

has measure zero. Then

$$\lim_{N \to \infty} \frac{1}{N} \nu(a, b, N) = \frac{1}{2\pi} \Omega(a, b).$$

The proof hinges on the fact that the conclusion of the corollary is exactly the conclusion of Theorem 6 with

$$F(x) = \begin{cases} 1, & a \le x \le b, \\ 0, & \text{otherwise.} \end{cases}$$

The reason the corollary does not follow immediately is that F is not continuous, and our task is to show that the result for continuous F can be extended to hold for at least this special discontinuous F under the assumption made.

Given a positive integer n, let G_n be the trapezoid-shaped function which is 1 in the interval $\left[a + \frac{1}{n}, b - \frac{1}{n}\right]$, 0 outside the interval $[a, b]$, and linear in the two intervals $\left[a, a + \frac{1}{n}\right]$ and $\left[b - \frac{1}{n}, b\right]$. Then each G_n is continuous and

$$\lim_{n \to \infty} G_n(x) = F(x)$$

except when $x = a$ or b. Now, since $G_n \le F$,

$$\frac{1}{N} \nu(a, b, N) = \frac{1}{N} \sum_{i=1}^{N} F(\lambda_i) \ge \frac{1}{N} \sum_{i=1}^{N} G_n(\lambda_i),$$

so by the theorem

$$\liminf_{N \to \infty} \frac{1}{N} \nu(a, b, N) \ge \frac{1}{2\pi} \int G_n[\phi(\theta)]\, d\theta.$$

This holds for each n, and the left side of the inequality is independent of n. Therefore the inequality still holds if the right side is replaced by its limit as $n \longrightarrow \infty$. But (and this follows from the Lebesgue bounded convergence theorem and our assumption concerning ϕ)

$$\lim_{n \to \infty} \int G_n[\phi(\theta)]\, d\theta = \int F[\phi(\theta)]\, d\theta = \Omega(a, b).$$

Therefore

$$\liminf_{N \to \infty} \frac{1}{N} \nu(a, b, N) \ge \frac{1}{2\pi} \Omega(a, b).$$

The inequality

$$\limsup_{N\to\infty} \frac{1}{N}\,\nu(a, b, N) \leq \frac{1}{2\pi}\,\Omega(a, b)$$

is proved similarly, and the two combine to give the result.

The second corollary we want to mention deals with the *determinants* of finite Toeplitz matrices. Let $D_{\phi,N}$ denote the determinant of $\mathbf{T}_{\phi,N}$. Since $D_{\phi,N}$ is the product $\lambda_1 \cdots \lambda_N$ of the eigenvalues, we have

$$\log D_{\phi,N} = \Sigma \log \lambda_i,$$

and the connection with the theorem is now clear.

COROLLARY 2: *If $m > 0$, then*

$$\lim_{N\to\infty} D_{\phi,N}^{1/N} = e^{(1/2\pi)\int \log \phi(\theta) d\theta}.$$

The expression on the right is known as the *geometric mean* of ϕ (just as $(2\pi)^{-1}\int \phi(\theta)\,d\theta$ is called the arithmetic mean of ϕ), and this corollary says that $D_{\phi,N}$ is approximately the Nth power of the geometric mean of ϕ. The proof is easy. Since $m > 0$ the function $\log x$ is continuous on $[m, M]$ and the theorem says

$$\lim_{N\to\infty} \frac{1}{N}\,\Sigma \log \lambda_i = \frac{1}{2\pi}\int \log \phi(\theta)\,d\theta.$$

The result now follows upon exponentiation of both sides.

5. REMARKS AND REFERENCES

Doubly Infinite Toeplitz Matrices. The main work of Toeplitz on the subject is the paper [11]. Toeplitz assumed that $\phi(\theta)$ was the value at $e^{i\theta}$ of a function analytic in a ring containing the unit circle, which is quite special. He did this since he used Laurent series rather than Fourier series. In fact Toeplitz called the associated quadratic forms *L-forms*, the *L* standing for Laurent.

Semi-Infinite Toeplitz Matrices. The first theorem of consequence relating to inversion of semi-infinite Toeplitz matrices is Theorem 3″, which is due to Wintner [13]. Shortly after this, Wiener and Hopf discovered a procedure for solving certain equations, which are now called Wiener-Hopf equations. (A presentation of this can be found in Chapter IV of the book [8] of Paley and Wiener.) These are integral equations of the form

$$\int_0^\infty K(x-y)f(y)\,dy = f(x), \qquad x \geq 0,$$

which are just the continuous analogues of the semi-infinite Toeplitz matrix equations

$$\sum_{n=0}^{\infty} c_{m-n}a_n = a_m, \qquad m \geq 0.$$

Thus, although we have been considering slightly different equations, the analogy is clear. The main device of the Wiener-Hopf technique is a certain factorization which is analogous to the factorization $\phi = \phi_-\phi_+$ we used in the proof of Theorem 5. Needless to say, one uses Fourier transforms for Wiener-Hopf equations just as one uses Fourier series for Toeplitz matrices.

Theorem 4 is due to Hartman and Wintner [5]. Theorem 5 was a highly cooperative affair. It was proved under the assumption $\Sigma\,|c_n| < \infty$ in the two independent papers [6] and [1], was proved in more (although not yet complete) generality in [12], and finally appears in general form in the paper [2] of Devinatz. This last paper, incidentally, presents the theory of generalizations of semi-infinite Toeplitz matrices to what are known as *Dirichlet algebras*.

The generalized Neumann series used in the proof of Theorem 2 can be found in any of the standard works on functional analysis; for example, Theorem 22A of [7].

Finite Toeplitz Matrices. The set filled in by the spectrum of $\mathbf{T}_{\phi,N}$ for a trigonometric polynomial ϕ has been determined by Schmidt and Spitzer [9]. This case, and the Hermitian case, are all for which the answer is known.

In 1958 there appeared the book [3] of Grenander and Szegö, which contains an enormous amount of material on finite Toeplitz matrices, their generalizations, and their applications. Szegö's original proof of Theorem 6 is given in Sec. 5.2. Curiously enough, Corollary 2, involving the determinants, is proved first, and the general theorem is derived from this special case.

REFERENCES

1. Calderón, A., F. Spitzer, and H. Widom, "Inversion of Toeplitz matrices," *Ill. J. Math.*, vol. 3 (1959), pp. 490–498.
2. Devinatz, A., "Toeplitz operators on H^2 spaces," *Trans. Amer. Math. Soc.*, 112 (1964) 304–317.
3. Grenander, U., and G. Szegö, *Toeplitz Forms and Their Applications.* Berkeley, Univ. of California Press, 1958.
4. Hardy, G. H., and W. W. Rogosinski, *Fourier Series.* Cambridge, Cambridge Univ. Press, 1944.
5. Hartman, P., and A. Wintner, "The spectra of Toeplitz's matrices," *Amer. J. Math.*, vol. 76 (1954), pp. 867–882.
6. Krein, M. G., "Integral equations on a half-line whose kernel depends on the difference of its arguments," *Uspeki Mat. Nauk,* vol. 13 (1958), pp. 1–120 (Russian).
7. Loomis, L. H., *An Introduction to Abstract Harmonic Analysis.* New York, Van Nostrand, 1953.
8. Paley, R. E. A. C., and N. Wiener, *Fourier Transforms in the Complex Domain,* Amer. Math. Soc. Coll. Pub. 19 (1934).
9. Schmidt, P., and F. Spitzer, "The Toeplitz matrices of an arbitrary Laurent polynomial," *Math. Scand.*, vol. 8 (1960), pp. 15–38.
10. Titchmarsh, E. C., *The Theory of Functions.* London, Oxford University Press, 1939.
11. Toeplitz, O., "Zur Theorie der quadrischen und bilinearen Formen von unendlichvielen Veränderlichen," *Math. Ann.*, vol. 70 (1911), pp. 351–76.
12. Widom, H., "Inversion of Toeplitz matrices II," *Ill. J. Math.*, vol. 4 (1960), pp. 88–99.
13. Wintner, A., "Zur Theorie der beschränkten Bilinearformen," *Math. Z.*, vol. 30 (1929), pp. 228–282.

INDEX